METHUEN'S MONOGRAPHS
ON CHEMICAL SUBJECTS

<hr>

General Editors

R. P. Bell, M.A., HON. LL.D., F.R.S.
N. N. Greenwood, D.SC., PH.D., SC.D.,
R. O. C. Norman, M.A., D.PHIL.

Elementary Molecular Bonding Theory

Peter G. Perkins

graham

B.SC., PH.D., D.SC.

Professor of Inorganic Chemistry
University of Strathclyde

METHUEN & CO LTD
11 NEW FETTER LANE, LONDON EC4

First published 1969
© Peter G. Perkins 1969
Printed in Great Britain by
Butler & Tanner Ltd, Frome and London
SBN 416 42680 8

Distributed in the U.S.A. by
Barnes & Noble Inc.

Contents

Preface

The basic theory underlying chemical bonding is becoming increasingly required by students and practising chemists. Fortunately it is not difficult to acquire a working knowledge of the elegant and powerful notions which have been developed and applied to the problem in recent years. Indeed, when stripped of their formal language, the essential ideas are simple enough to be presented to the first-year student or the senior sixth-form pupil.

This book is devoted to bringing to an elementary level some of these ideas and to showing them at work in chemistry. An attempt has been made to avoid 'lies' which would have to be 'unlearnt' at a later stage in the pupil's development. The alternative approach has been to present relatively advanced ideas in high dilution.

The book should be suitable for reading by senior sixth-form pupils and first and subsequent year degree students in Universities and Technical Colleges.

PETER G. PERKINS

CONSTANTS

Electronic charge, e $= 4 \cdot 8030 \times 10^{-10}$ esu $=$
                                  $1 \cdot 6021 \times 10^{-19}$ coulomb

Electron mass, m $= 9 \cdot 1090 \times 10^{-28}$ g

Proton mass, Mp $= 1 \cdot 6722 \times 10^{-24}$ g

Velocity of light *in vacuo*, c $= 2 \cdot 9979 \times 10^{10}$ cm sec^{-1}

Planck's constant, h $= 6 \cdot 6255 \times 10^{-27}$ erg sec

Bohr radius, a_0 $= 0 \cdot 529$ Å

CONVERSION FACTORS

1 Ångström (Å) $= 10^{-8}$ cm $= 0 \cdot 1$ nm

1 atomic unit (au) $= 0 \cdot 529$ Å $= 0 \cdot 0529$ nm

1 electron volt (eV) $= 8{,}066$ cm^{-1} $= 23 \cdot 06$ kcal mole^{-1}
                                              $= 9 \cdot 649 \times 10^4$
                                                   J mole^{-1}

CHAPTER 1

Valency

Chemistry is a fascinating subject because it surrounds us all the time and governs our daily lives. Rocks, plastics, alloys, bacteria, trees, birds' feathers; all are built up from a strictly limited number of chemical elements combined together in an infinite variety of ways. Indeed, for living materials only a small number of chemical elements are required (e.g. C, H, O, N, S, P) although there are some ninety to choose from. The chemical combinations which we call molecules result from chemical reactions favoured by energy factors. This is the dynamic aspect of chemistry and is a topic which is, of course, of extreme importance but one which we shall not treat in this book. It is the fact that elements do join together as molecules in such a vast number of combinations and variety of shapes which is the most intriguing. Our curiosity is excited by the often highly symmetrical arrangement of regulated numbers of atoms around others and by the diversity of problems this raises.

It is the purpose of this book to enquire why atoms combine with each other, to obtain and use the rules which govern combination and to find a comprehensive approach which will explain and describe the *bonds* which hold atoms together in molecules. We should then be able to see why certain arrangements are found and perhaps to predict possible new ones. It is impossible, in so short a space, to elaborate in detail the profusion of ideas which have been advanced to accomplish the above ends. We shall find it better to pose a number of pertinent questions and to discuss the answers in as consistent and simple a manner as possible. The subject matter will be mainly devoted to explanation and use of the modern approach to chemical bonding. This will be, in the main, non-mathematical but the author wishes to state here the belief that frequently the easiest and quickest way to grasp the concepts in chemical theory is to study the simple mathematics involved.

Any elementary approach to chemical theory should be, as far as possible, free from deliberate errors (even white lies!) which will require 'unlearning' at a later stage. There is no reason why correctness should not be achieved as the elements of the most up-to-date theory are intrinsically simple and elegant and can be studied at an early stage. Later these ideas may be extended and elaborated upon and the technical aspects (the mathematics) brought in to make the approach more quantitative.

The Problems of Bonding and Shape

What then are the questions we have to answer? Firstly, why do atoms bond together to form molecules instead of being 'content' to remain separated? Why do some others, e.g. helium, neon, argon not form molecules? Any explanations advanced for bonding must account for the latter observation too. Secondly we must explain why the atoms in molecules cluster together in fixed proportions, e.g. why do only two atoms associate to form a molecule of hydrogen? Why cannot a third or a fourth approach and attach themselves to the first two? How does the phenomenon of saturation (of which this is an example) arise? This and other questions like it will not be answered directly but will be treated as problems to be approached open-mindedly and the answers *derived*. Thus the philosophy will not be to try and assign *a priori* valence number to atoms, but will be to develop a number of fundamental principles so as to facilitate the treatment of a molecular model on its merits. Thus molecules such as H_3 and H_4 will be treated just as seriously as will H_2 itself.

Thirdly, what are the forces which hold atoms together – in other words what is a chemical bond? Why do certain compounds give rise to ions when they are placed in an ionizing medium such as water? What is the bonding between the atoms in such compounds – are they held together in a different way to the rest of chemical compounds?

Fourthly, why do atoms associate as molecules not only in set proportions but also in certain definite geometrical shapes. In short, how does this mysterious property possessed by atoms called 'valency' arise?

The Early Development of the Theory of Atomic Structure

The first step in the investigation is to consider the principal ideas concerning the structure of chemical compounds which held sway before the advent of present views. Dalton, in his Atomic Theory, laid the foundation of the combining laws for atoms, i.e. the science of stoichiometry; much other work during the same period, by Gay Lussac and Berzelius, was devoted towards similar ends. The genius of this work can only be properly realized when one remembers that, at the time, few physical laws were recognized (and indeed those that were dealt with bulk properties of systems only). It goes without saying that no chemical structures had been elucidated, and only the vaguest theories of atomic structure had been proposed. Of course, since Dalton assumed that atoms were hard, smooth spheres he could not suggest why they should apparently be attracted to each other. Later work on the conductivity and electrolysis of solutions of certain substances in water suggested that they dissociated into conducting particles (or ions) of opposite charge. However, this still did not provide an answer to the problem of what constituted a chemical bond although it had been suspected for some time that interatomic forces were electrical in nature.

Perhaps the most significant advance which eventually led to an explanation of chemical bonding was made about a century later. It was the culmination of three, apparently unrelated, advances in physics. Firstly the *atomic emission spectrum* of hydrogen was obtained. The emission spectrum of hydrogen results when the element contained in a tube is subjected to an electric discharge. Energy is taken in and subsequently the gas emits this energy as light. The results are recorded on a photographic plate, and it is discovered that the light is emitted only at certain definite frequencies. Thus the atomic emission spectrum of hydrogen consists of a set of lines on the plate and each line corresponds to a particular frequency at which light is emitted. Next, Rutherford and Moseley showed that atoms were composed of a central, positively charged nucleus (wherein most of the mass of the atom resided) surrounded by extranuclear electrons. The number of units of positive charge contained in the nucleus had to be equal to the number of electrons because the whole was, overall, electrically neutral. Meanwhile concurrent research by Max Planck along different lines led to his revolutionary suggestion (1900) that energy cannot be absorbed or radiated continuously, but only in small 'lumps' or *quanta*. This idea may seem difficult to understand at first because, for example, we do not move about in jerks nor do we feel short bursts of heat from an electric fire, but it should be remembered that water runs continuously from a tap and yet we know that it is composed of very small lumps of matter (molecules). Alternatively, if we look at a newspaper photograph with the naked eye it appears to be quite continuous but on studying it more closely through a magnifying glass it will be seen to be made up of a large number of small dots. It is important to appreciate the meaning of the Quantum Theory because it forms the essential basis of all the development which followed.

The experimental observation of the hydrogen spectrum seemed to be connected in some way with both the single extranuclear electron of hydrogen and the Quantum Theory. Finally, in 1913 Bohr coordinated the ideas and results of all three researches and put forward a theory to explain both the atomic spectrum of hydrogen and the stability of the atom. With regard to the latter problem, Rutherford had suggested that, in atoms, the negative electrons revolved around the positive nucleus. Since it was supposed that these both must be particles it was natural to try to describe the atom in terms of classical mechanics which embodied well-established theorems and applied to large bodies. The difficulty here however is that, on the basis of classical mechanics, a moving charge must radiate energy and so a circling electron would lose momentum and eventually spiral into the nucleus. To overcome this objection Bohr made the drastic assumption that electrons around a nucleus did not have to radiate energy, i.e. they occupied stationary states and their energy was conserved whilst they stayed in these states. This fixed path of the electron around the nucleus was circular and was called an orbit, and there were a number of definite orbits associated with different energies in which the electron could travel. It is worthwhile to study the Bohr Theory in a little more detail because of the impetus it gave to the eventual development of modern theory.

Let us first see how well it fits the hydrogen spectrum. The physicist Balmer had shown, without the aid of any theory, that the frequencies of many of the lines were interrelated in a simple way, thus

$$\nu \propto \frac{1}{2^2} - \frac{1}{n^2}$$

where ν is the frequency of the line and $n = 3, 4, 5 \ldots$ The Balmer formula turned out to be merely a special case and it was shown that *all* the lines in the hydrogen spectrum could be obtained from a simple generalization of the above formula so that

$$\nu \propto \frac{1}{n_1^2} - \frac{1}{n_2^2}$$

where $n_1 = 1, 2, 3 \ldots$ and $n_2 = 2, 3, 4 \ldots$ (for $n_1 = 1$); $3, 4, 5 \ldots$ (for $n_1 = 2$); and $4, 5, 6 \ldots$ (for $n_1 = 3$). The second of these sets of numbers corresponds to the special case of the Balmer series. Bohr assumed that the hydrogen orbits were characterized by a number n, called the *principal quantum number*, and whilst the electron remained in a particular orbit its energy was fixed. However, it could absorb or emit energy and 'jump' to another orbit. Since the orbits were the only possible paths in the atom for the electron, in order to change orbit it had to absorb or emit a 'burst' of energy. It should be recognized that this is where quantum theory comes into force and implies the electron can only have a set of *quantized* energies. The hydrogen emission spectrum can now be readily explained using the Bohr model of the atom and the quantum theory. Each line in the hydrogen spectrum is caused by the emission of a quantum of energy (as light). The energy corresponds to a single, sharp frequency of light and the electron moves from one orbit to another. For a hydrogen atom to emit light it must of course first absorb energy. Absorption can be effected either by an electric discharge or by irradiating the element with light at a series of different frequencies. As with emission the light is absorbed only at certain frequencies, and so as the light shining on the hydrogen atom is varied in frequency there will arise certain frequencies which correspond to energy separations between the orbits in the atom. Light will be absorbed at these, and only these, frequencies and will cause the electron to move out of its orbit to another. Figure 1.1 illustrates the principles with the Balmer series as an example. The Balmer series of spectral lines arises from jumps by the electron to or from the $n = 2$ level to the levels for which $n = 3$, 4, 5, etc. The frequencies of light emitted or absorbed are proportional to each other as shown by the Balmer form of the general equation. The frequencies of lines in other series where n_1 is not equal to two may be calculated by the general formula and moreover have been discovered experimentally. In conclusion Bohr was, in fact, able to derive the general formula for line frequencies from first principles.

The Bohr Theory was a major breakthrough in understanding atomic structure but further modifications were to come. Closer examination revealed fine structure in the emission spectra of

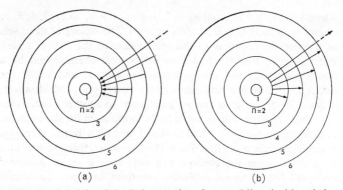

FIGURE 1.1. Origin of the Balmer series of spectral lines in (*a*) emission, (*b*) absorption

heavier elements – what looked like single spectral lines turned out to be closely spaced groups of lines. In explanation Sommerfeld postulated that the Bohr circular orbits were just a special case. In general, atomic orbits should be elliptical. The ellipticity of the orbit could then be characterized by a further quantum number, k which took the values $1 \dots n$. This number was later changed to l; it had values $0 \dots n - 1$ and was called the *azimuthal* or *angular momentum* quantum number (remember n has an unlimited range of values). This means that each n number governs a *group* of l numbers and different l numbers are associated with slightly different energies, e.g. the orbit $n = 2$, $l = 0$ had slightly different energy to that with $n = 2$, $l = 1$. Hence an electron jumping from another level could produce two spectral lines because it could terminate in either of these sub-levels, the energy of the two jumps being slightly different.

The next discovery was that if the hydrogen which was emitting light according to the above rules was placed in an *intense magnetic field* then the observed lines were split up into more hyperfine constituents. This implied that there were yet more sub-levels in the system. The new lines were easily explained by assuming that each l number of a n level was *itself* associated with a set of levels defined by a *magnetic quantum number m*. This could take the values, $-l, (-l + 1) \ldots 0 \ldots (l - 1), l$ (i.e. $2l + 1$ values). Almost unbelievably, in 1926, Uhlenbeck and Goudsmit, examining the atomic spectra of alkali metals, discovered yet more fine structure; some hyperfine lines turned out to be split into two. The splitting was attributed to the *spin* of the electron, i.e. an electron in an orbit could spin on its own axis in one of two ways, i.e. either parallel or antiparallel to its orbital motion. The spin was governed by a quantum number s or m_s which could be equal to either $+\frac{1}{2}$ or $-\frac{1}{2}$. No one could say why the numbers $\pm\frac{1}{2}$ were appropriate – they just happened to fit the spectroscopic data.

To sum up: an electron in an orbit could be characterized by four quantum numbers which were

  n the principal quantum number – governing the *size* and *energy* of the orbit,

  l the azimuthal or angular momentum number – governing the *ellipticity* of the orbit,

  m the magnetic quantum number – governing the splitting of the l levels in a magnetic field (or the orientation of the orbit in a magnetic field),

  s the spin quantum number, equal to $\pm\frac{1}{2}$, and governing spin orientation.

Later, in order to explain the atomic spectra of heavier elements, Pauli suggested the principle that *no two electrons in an atom could have the same four quantum numbers*. This means that two electrons could only be characterized by the same n, l and m if one of them had $s = +\frac{1}{2}$ and the other $s = -\frac{1}{2}$. The maximum number of electrons per orbit was therefore limited to two.

The Electronic Theory of Valency

The Bohr Theory provided a sound starting-point for the development of theories of bonding between atoms and these were indeed quick to arrive. For some time previous to the Bohr Theory of the atom, in fact since Thomson's discovery of the electrically charged electron in 1897, chemists suspected that electrons played a pivotal part in the mutual bonding of atoms – a supposition which correlated well with the electrical forces known to operate in materials dissociable into ions in solution. It was also quite well recognized at that time that there were two types of compounds – the polar and the non-polar. The earliest *definitive* work on chemical bonding was carried out simultaneously and independently in 1916 by Kossel and by Lewis and Langmuir. The whole became known as the *electronic theory of valency*.

Bonds between atoms were supposed to be of two types: firstly, the *electrovalent* type, in which one of the participating atoms lost one or more electrons to the other atom or group. The resultant positive and negatively charged entities then exerted Coulombic attraction upon each other. This was the essence of the bonding in *ionic* materials such as the sodium, potassium or magnesium salts. The *electrovalency* of an atom was defined as the number of electrons given up or gained by it. Such electrovalent compounds were found generally to be formed between the Group I elements (the alkali metals) and Group VII elements (the halogens). These were known (from their atomic numbers) to possess one more and one less electron respectively than the inert gases next to them in the Periodic Table. This led to the assumption that the outer configuration of the inert gases (an octet of electrons, except for He) was particularly stable and so, e.g. when sodium and chlorine formed an electrovalent bond the sodium and chlorine atoms both achieved an outer octet of electrons by forming Na^+ and Cl^-. This favourable situation was the main reason for the formation of the bond and the formulation led easily to a satisfying explanation of the conductivity of aqueous solutions of such compounds and Faraday's Laws of electrolysis. The concept of inert gas stability was also instrumental in advancing the theory of bonding in non-polar compounds. Lewis proposed that bonds in non-polar systems were formed by the *sharing* of electron pairs rather than transfer of electrons from one atom to another. Thus the concept of the electron-pair bond originated and was rapidly subject to considerable improvement and extension.

Electron-pair bonds later became divided into two types:

(i) Covalent bonds

In these situations each atom was considered to contribute one electron to the bond between them. Covalency is well exemplified by the fluorine and hydrogen molecules (Figure 1.2). In the former

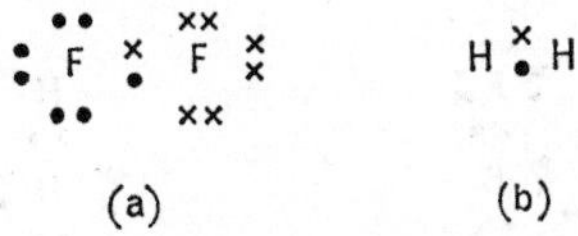

(a)   (b)

FIGURE 1.2. Kossel–Lewis representation of the (a) fluorine and (b) hydrogen molecules

both fluorines have one unpaired electron in their outer shells and these pair up so that each fluorine atom takes a share in the *pair* of electrons. This results in an octet of electrons around each fluorine and so, it was supposed, the system was stable. The hydrogen case is a trifle different. There are insufficient electrons to form octets and stability was considered to be due to the achievement of the helium configuration by each hydrogen atom.

The electron-pair notion of covalent bonding gave much support to the idea of *valency* or '*combining power*' being an atomic property interpreted in terms of the number of bonds to other moieties which the atom will form. It is our purpose to review this idea, and eventually to revise it.

(ii) The coordinate bond

This was still essentially an electron-pair bond arising from combination of two molecules wherein one provided *both* electrons for the bond. The classic example of this type of bond is the compound $BF_3.NH_3$: BF_3 itself may be formulated in a similar way as for F_2 or H_2 (Figure 1.3 (a)) with simple electron-pair covalent bonds, but it differs from fluorine in that the boron atom does not gain an octet by bonding to three fluorine atoms. However, ammonia

B

(formulated similarly, Figure 1.3 (*b*)) possesses a 'lone-pair' of electrons in its octet and so when the two molecules are joined together (Figure 1.3 (*c*)) all octets (or duets) are completed. Other names have been given to this type of bond: semipolar, dative-covalent, co-ionic, donor-acceptor, but we shall always refer to it as the coordinate bond.

The ideas of Kossel and of Lewis and Langmuir are simple, straightforward and easy to apply. It is not surprising that they held almost complete sway in textbooks until relatively recently; indeed

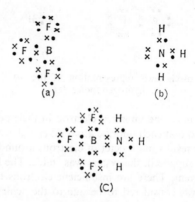

FIGURE 1.3. Kossel–Lewis representation of (*a*) boron trifluoride, (*b*) ammonia, (*c*) the $BF_3 . NH_3$ coordination complex

even in some advanced areas of inorganic chemistry, disguised in more sophisticated language, they still do. There is no doubt that they embody the essential truth; electron-pair bonds can endow chemical systems with great stability. The same idea carries through into modern theories although in a somewhat modified form: bonds need not necessarily involve pairs of electrons but it is usually advantageous if they do.

The impression that *only* an octet of electrons can confer stability is, however, less useful and is demonstrably incorrect. An example of this has already been encountered in the planar compound, BF_3, where the boron atom is formally surrounded by only six electrons. This, and other, trigonal boron compounds with similar electronic formulation are not *intrinsically* unstable, e.g. they do not decompose into their elements readily on heating. It *is*, however, true that many of them are able to 'accept' a pair of electrons from an electron-rich source (such as nitrogen in NH_3) and so 'complete the octet'.

More well-defined examples are, however, easy to find: the compound sulphur hexafluoride (SF_6) is extremely stable both towards heat and chemical attack. Unless this compound is considered to be ionic with six F^- ions surrounding a central S^{6+} ion (unlikely as it is a non-conductor of electricity) it is clear that the sulphur atom cannot be surrounded by eight electrons; the number is, in fact, formally twelve. A further example is provided by the ferrocyanide ion, $Fe(CN)_6^{4-}$ (this should be familiar as one reactant in the Prussian-blue test for ferric iron in qualitative analysis). In this unit each CN^- ion forms a coordinate bond to the central Fe^{++} ion so the latter is surrounded by at least twelve electrons. It should be emphasized that this ion is not unusual or exceptional in its electronic structure but is only one of thousands of similar ions and compounds of the so-called transition elements. We conclude that the Lewis octet rule works best for elements in the main groups of the Periodic Table, particularly those only one or two groups removed from the inert gases. Furthermore, the desirable situation wherein all two-atom bonds are associated with two electrons cannot always be realized in compounds and when it cannot the Kossel–Lewis–Langmuir approach is uninformative. Hence it represents merely a special case which can be incorporated in a much more general theory and it is this that we shall try to develop.

The stability of a molecule is attributed to the chemical bonds therein and these bonds have been classified as ionic, covalent and coordinate. The strength of an ionic bond resulted from electrostatic attraction between the positive and negative ions. This is readily understandable. On the other hand the strength of the covalent and coordinate bonds was ascribed to the *pairing* of electrons between atoms, in the latter case both being provided by the same atom. There is an obvious difficulty in understanding the covalent bond; what is magical about electron pairing (e.g. in H_2) which makes pairs of hydrogen atoms more stable than isolated ones? Furthermore, chemists recognize that some bonds are not completely covalent but have some degree of ionic character. How should we explain their stability? Is there a gradual transition from covalent to ionic character over a series of bonds or is there a discontinuous change of bond type at some point? At what stage does electrostatic attraction stop and 'electron-pairing forces' take over? There should surely be some simple connection between the two because, after all, they are both due to electrons.

Difficulties arising from the Electronic Theory of Valency

A natural outcome of the notion of the electron pair bond was to represent each of these in a molecule by a line (or 'stick') joining the atoms involved. The trouble with this is that not all bonds in molecules (even some simple ones) are associated with a pair of electrons and it is of course not possible to write non-integral numbers of 'sticks'. This leaves many perfectly respectable bonds incompletely represented. We shall be studying some examples of bonds of this kind in later chapters. The Kossel–Lewis Theory was linked in an

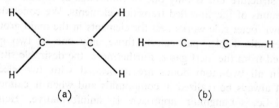

(a)   (b)

FIGURE 1.4. Structures of (a) ethylene and (b) acetylene

obvious way to the observed stoichiometry of molecules and the *valency* of the atom. One definition of the valency frequently used at the time was the number of hydrogen atoms with which the element would combine. This implied that the valency coincided either with the number of the Periodic group or with the number '8 − n', where n is the Periodic group number.

It is worth analysing this definition a little further. In methane, CH_4, a carbon atom combines with four hydrogen atoms from which it might reasonably be concluded that the valency of carbon is four. But in ethylene, C_2H_4, which has the geometrical structure shown in Figure 1.4 (a), each carbon is combined with two hydrogens and one other group, a second —CH$_2$ unit. Hence to be consistent we should say that in this compound the valency of carbon is three and likewise in acetylene (Figure 1.4 (b)) the valency of carbon would be two. On this basis therefore even a familiar element like carbon has at least three valencies. The problem may be circumvented fairly easily by modifying the definition of valency. Suppose it is redefined as the *number of bonds* that the atom can form to other atoms or groups. Thus in ethylene the C—C part should be written as C=C, i.e. a double bond, and in acetylene there should be a triple bond between the two carbon atoms. All is now satisfactory because the number of bonds formed by carbon in each case is four and this becomes a valency number attributable to carbon and usable in prediction of the type of compounds which might be formed by this element. At this juncture it should be noticed that there is an element of circularity in the argument because the double and triple bonds in C$_2$H$_4$ and C$_2$H$_2$ were drawn mainly to satisfy the assumed quadri-valency of carbon!! Luckily, in these particular instances, independent evidence for the double and triple bonds exists. Use of the valency number four as an immutable quantity for carbon does indeed turn out to be extremely useful and can rationalize and coordinate most of the structural aspects of organic chemistry.

Difficulties immediately arise if we try to apply the same rigid reasoning to inorganic compounds. In sodium chloride vapour a single molecule has 1 : 1 stoichiometry (Na : Cl) and by previous arguments there is a single electrovalent bond between the two. This would be represented by a single 'stick' making the valencies of both sodium and chlorine unity. But let us look more closely at this case. In the solid state (as usually encountered) sodium chloride does not contain isolated NaCl molecules but is an ionic, three-dimensional, infinite network (Figure 1.5). Now we cannot specify to which chloride ion a particular sodium ion is bonded, it is surrounded by six of these and, of course, the same reasoning applies equally to the chloride ions. A sodium ion (e.g.) attracts each of the six chloride ions equally and indiscriminately and must be bonded to each. Hence Na$^+$ and Cl$^-$ each form six bonds in NaCl crystal and by the defined criterion of valency, a valency of six must be assigned to these elements. But for a molecule of the same compound we have just argued that the valency is unity. What is the answer?

It is not satisfactory to say that each bond in the crystal is only one-sixth of a 'full' ionic bond – we did not specify any requirement of strength in the interactions we consider to be bonds. Should we

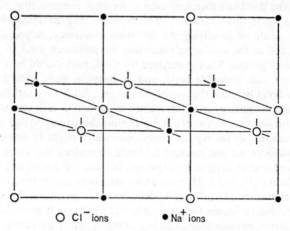

O Cl⁻ ions   • Na⁺ ions

FIGURE 1.5. Section of the sodium chloride lattice

reject the definition of valency and not count up the number of bonds when they are ionic, i.e. should a different definition be adopted when electrovalency is being discussed? If so, what will happen when we meet bonds which are partially ionic? These questions are left unanswered for the time being; the answers will emerge later.

The next case of interest is ammonium chloride, which is produced by the reaction of ammonia and hydrogen chloride. The NH_3 molecule has already been considered (Figure 1.3 (*b*)) and under the '8 − *n*' rule a valency of three would be assigned to the nitrogen atom because it forms three covalent bonds. The ammonium ion in ammonium chloride is like methane in shape, i.e. it consists of a central nitrogen surrounded tetrahedrally by four hydrogen atoms. The *whole NH₄ unit* has one formal positive charge and the Lewis pattern of the ion is depicted in Figure 1.6.

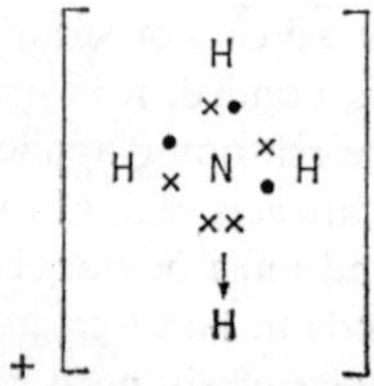

FIGURE 1.6. Kossel–Lewis representation of the ammonium ion

The nitrogen atom is bonded to the four hydrogens by three covalent bonds and one coordinate bond. What is the valency of nitrogen in ammonium chloride? Depending on the definition of valency adopted the answer could be 3, 4 or 5!!

The valency of N might be considered to be three, because NH_4^+ incorporates three covalent bonds to three hydrogens, or four, because in addition there is a fourth *coordinate* bond to the fourth H^+ ion. We cannot simply say that the latter bond does not count because (*i*) it was agreed initially to count covalent bonds (that is how the number 4 is obtained for carbon in all the organic compounds) (*ii*) since the NH_4^+ ion is regularly tetrahedral, like methane, the three 'covalent' and the one 'coordinate' bond cannot be inter-distinguished. If one is discounted, so must be all the others! Thus, on this reasoning, the valency of nitrogen is 4 (or zero, if one took the pessimistic view). Now in an isolated NH_4Cl molecule, there is a further ionic bond between the NH_4^+ unit and the Cl^- ion, just as in NaCl. We cannot neglect this either because then, by identical logic, we would have to discount the ionic bond in sodium chloride also. An additional valency of nitrogen has thus arisen, making the whole up to five. Unfortunately, this is not the last word; we can, without any trouble, complicate the issue still further. To begin with, as in NaCl crystal, each ammonium ion attracts *all* its nearest negatively charged neighbours. Secondly, the unit positive charge of NH_4^+ *does not reside solely on the nitrogen atom* but is distributed over all the constituent atoms. The electron transferred to the Cl atom does not originate from any particular atom of the NH_4 moiety and so the electrovalency cannot be ascribed to either nitrogen or hydrogen. This upsets the initial concepts because it is somewhat hard to fit in the 'chlorine-attracting character' of the hydrogen atoms. Should it be considered as a 'secondary valency'?

Difficulties of this kind are not restricted to NH_4Cl. An isolated water molecule is angular (Figure 1.7 (*a*)) from which we would say that oxygen is divalent and hydrogen univalent. However, in a crystal of ice, hydrogen bonding occurs (Figure 1.7 (*b*)) and the oxygen is *tetrahedrally* surrounded by four hydrogens, two of which are nearer than the other two. The two hydrogens nearest to the oxygen atom are covalently bonded to it and the two remote from it are bonded electrovalently. The latter attraction stems from small, induced, positive and negative charges on the oxygen and hydrogen atom respectively. By our new reasoning, i.e. counting up bonds of any strength, we would have to say that the valency of oxygen in ice is four and that of hydrogen two – this way our definition is satisfied. The overall problem is not yet finished though; going back to the case of solid sodium chloride, the hexa valency of Na⁺ was suggested on the grounds that the species is ionically bonded to six

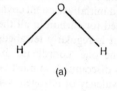

(a)

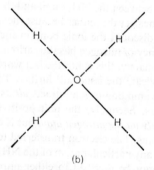

(b)

FIGURE 1.7. Structures of H_2O

nearest Cl^- ions. But it is also attracted, more weakly to be sure, to other non-adjacent Cl^- ions. Should we also count these attractions as 'valencies'? It is no way out to reply that such attractions are weak and can therefore be discounted because this attitude begs the question: 'How strong do bonds have to be before they are considered as bonds – in other words, when is a bond not a bond?' The answer is surely that we cannot draw an arbitrary line beyond which interactions will be ignored and so a closed solution to the problem is not attainable along these lines.

The state of affairs now created is indeed a puzzling one: it seems that we can only continue to talk about the valency of an element if the environment and the bonds considered are both specified. This means treating every case on its merits. In inorganic chemistry we are almost inevitably led to the conclusion that there is no well-defined number to assign to an atom and which is usable for all cases. On the other hand, in organic chemistry there are rules which allow systematization. There is no argument that one should neglect such rules but that it should be realized that they represent special cases where a rigid application of the valency criterion for carbon will yield the right answers. The valency dogma worked well enough at first in inorganic chemistry but later it almost certainly inhibited development in the understanding of new classes of compounds which did not conform to a set pattern of atomic valencies. These were thought to be 'anomalous' and were either set aside or special treatments were invented for them.

What we must now try to create is a unified approach to valency which operates not only for compounds of carbon, but also prepares the way towards an understanding of the, now vast, number of 'anomalous' or 'non-classical' compounds which have been prepared. Our approach, then, will be to develop principles which will allow construction of a scheme of bonding for any molecular geometry and by which we can appreciate the way in which it is held together. This is better than assigning a valency number to an atom and then trying to make the structure in question fit to it. In so doing we should recover the valency rule where this is obeyed and may also discover why one particular geometry of a compound is more stable than another, e.g. why methane has tetrahedral rather than planar geometry.

CHAPTER 2

Theory of Atomic Orbitals

The Bohr Theory of the atom was clearly a brilliant piece of thinking, opposing, as it did, the classical mechanical picture of physical systems and stating that the physical laws of macrostructures might well not apply to sub-atomic particles. However, it did have certain unsatisfactory features: (*i*) it applied well to the hydrogen atom (indeed, it was constructed to explain the emission of light of certain frequencies by hydrogen atoms) but it broke down badly when attempts were made to use it to explain the spectra of heavier elements. Perhaps more important is that (*ii*) the quantum numbers did not *arise* naturally but they and their interrelationships were *imposed arbitrarily* in order to get agreement with the experimental hydrogen spectrum. (*iii*) The electrons were considered to move in fixed paths, the orbits, from which they did not deviate unless a jump was made to a different orbit.

The next advance in the theory of the atom was attained by drawing an analogy between the nature of light and electrons. Important experiments such as the Compton and photoelectric effects, and wave diffraction and interference phenomena, had shown that light could exhibit both wave and particle properties and the form in which it acted depended on what experiment one carried out. Following on from this, in 1924, De Broglie suggested that with moving particles (such as electrons) there were associated waves and, moreover, the wavelengths were related to the mass of the particle (a corpuscular property) by the simple formula,

$$\lambda = \frac{h}{mV}$$

where λ is the associated *wavelength*, h is a constant (Planck's constant from Quantum Theory), m is the mass of the particle and V is its velocity. The latter two quantities together constitute the *momen-

tum of the particle. This equation is of fundamental importance because it provides a quantitative relationship between the wave and particle properties of a body.

For large bodies (which are the subject of classical mechanics) *m* is so great that λ becomes too small for experimental observation. Thus a tennis ball in 'service' has, in principle, wave properties, but no experiment can be set up to observe them. Large bodies therefore simply represent a limiting case of all particles and classical mechanics, which applies to them, is just one special branch of a mechanics which is much more general.

Next Heisenberg put forward his, now famous, 'Uncertainty Principle' which states that it is impossible to obtain *simultaneously*, precise (or *sharp*) values for both the momentum and position of a sub-atomic particle. This means essentially that, if we obtain a value for the momentum of an electron then we cannot trace its path. Since we are seeking to discover the momenta (and energies) of electrons in chemical systems we must accept the restriction of not knowing exactly their positions. This, luckily, does not matter much because alternatively we can discuss the *probability of finding an electron at a point or in a certain region*. Fairly obviously the probability of finding an electron is related to the *electron density* at that point or in that region. Probability is a rather abstract idea but electron density, or number of electrons at points in an atom or molecule, is something one is more able to grasp.

The Schrödinger Wave Equation

So far, all the reasoning has been rather qualitative but clearly the treatment *can* become quantitative because if there can be electron-waves then there should be a wave equation to represent their behaviour. In 1927, just such a wave equation was developed by Schrödinger. It has no simple derivation but in order to make further progress along the new lines we need to examine this fundamental Schrödinger equation. Its explicit form is

$$\frac{\partial^2 \psi}{\partial x^2} + \frac{\partial^2 \psi}{\partial y^2} + \frac{\partial^2 \psi}{\partial z^2} + \frac{8\pi^2 m}{h^2}(E - V)\psi = 0$$

where *E* and *V* and *m*, are the *total* and *potential* energies and the *mass* of the electron respectively, and *ψ* is a *wave function*. This equation need not engender any apprehension because we do not need to know anything at all about how to solve it, it *has* been solved and all that need interest us are the particular forms of the functions ψ which will satisfy it. The simplest cases to which the equation may be applied are hydrogen-like atoms. These are systems which possess a *single* electron outside the nucleus. Hydrogen itself consists of only one proton and one electron and the movement of the latter may be described if we first regard the nucleus as fixed. There is then no other moving particle which must be accounted for and the only influence on the single electron is the nuclear charge. The H-like system is the only one for which the wave equation can be solved exactly. This is a pity but it is not a fundamental weakness of the theory because (*i*) many qualitatively correct results may be obtained without any computation at all, (*ii*) approximate methods of solution for more complex systems exist and their precision is often only limited by our patience in using them.

Now we want to apply the equation to atoms which are spherical whereas the above form employs the variables, x, y and z which are more appropriate to 'square boxes' (e.g. rectangles, cubes, etc.). Accordingly it is usual to re-express and solve the equation in terms of three new variables (or coordinates) which fit the spherical geometry of the situation. These are r, θ and ϕ and they are called *polar coordinates*. There is a simple relationship between x, y and z and r, θ, ϕ which is illustrated in Figure 2.1.

Thus instead of specifying the point P in terms of distances along the x-, y- and z-axes we can pinpoint it by moving away from z by the *polar* angle, θ, around in the x, y plane by the *azimuthal angle*, ϕ, and then out in the resultant direction by the distance r. This is an alternative way of specifying points in space which is more appropriate to a sphere. It corresponds closely to latitude and longitude on the surface of the earth. Clearly if r is set constant and θ and ϕ varied arbitrarily then the point P would travel over the surface of a sphere of radius r.

If these three variables are now used in the wave equation instead of x, y and z the wave function, ψ, will be made up of r, θ and ϕ in some way. There is a standard technique for solving differential equations of this type and although the details of this are not particularly relevant to us here the results are. It is found that the form of a wave function is always the same, i.e. it is made up of a product of three *separate* functions each of which contains only one of the variables r, θ, ϕ. Thus a wave function in polar coordinates is,

$$\psi(r, \theta, \phi) = \underset{(1)}{f(r)} \times \underset{(2)}{f(\theta)} \times \underset{(3)}{f(\phi)}$$

Function 1, $f(r)$, contains *only* the radial variable r and because of this it is called the *radial function*. Of the three variables, r, θ, ϕ, the former is the only one which measures distance (Figure 2.1) and the function containing it is therefore the only one which can be correlated in any way with physical distances in an atom.

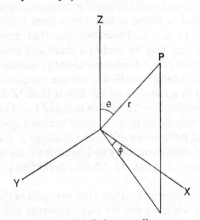

FIGURE 2.1. Polar coordinates

Functions 2 and 3, $f(\theta)$ and $f(\phi)$, both contain angular variables only and hence are usually combined and collectively called the *angular function*. The angular function incorporates no distance variable and so can never be related to physical distance.

We have been talking about the wave function as though there was only *one* form which would satisfy the Schrödinger equation. There are, in fact, whole families of functions which will do this and the *total* solutions for the atomic case are called *atomic orbitals*. Accordingly we shall now consider them in more detail as an understanding of bonding in molecules relies on a proper appreciation of their properties. It is convenient to examine them in turn.

The Radial Functions

Each radial function of the family has always the same form and is the product of two functions in the variable r,

$$f(r) \text{ (or } R(r)) = \text{constant} \times \text{(a simple polynomial in } r)$$
$$\times \text{(exponential term in } r)$$

The multiplying constants are called *normalizing constants* for the functions. They contain two integral numbers n and l and these have similar significance to that in Bohr Theory (see later for a more complete discussion of this point). Thus n is the *principal quantum number* and l the *azimuthal* or *angular momentum quantum number*. These are related as before in that n runs from 1 to n while l takes values from 0 to $n - 1$. These two quantum numbers are very useful because they may be used in a shorthand notation to designate the different radial functions. Usually, instead of specifying the l number of a radial function letters are substituted thus:

When $l = 0$ in a function we say that it is an 's'-function; when $l = 1$, a 'p'-function, $l = 2$ a 'd'-function and $l = 3$ an 'f'-function. Because l can only go up to $n - 1$ there can be a $2p$-function and a $3d$-function but not a $1p$- or a $2d$-. The choice of s, p, d, f in naming has traditional significance only and stems from the old description of atomic spectral series as '$s$harp', '$p$rincipal', '$d$iffuse', and '$f$undamental'.

It is now advantageous to study a few examples of radial functions. In this way we will get a feel for their properties and also tie in the nomenclature described above to their forms. The first few functions in the family are given in Table 2.1.

A regularity in the forms of these functions should at once be evident. The maximum exponent of the polynomial in r is r^{n-l-1} where n is the integral principal quantum number. Thus n exerts a powerful effect on the form of the radial functions and later it will be seen how this affects the energies of electrons described by these functions. We shall find out more about the functions if we plot them out graphically.

Table 2.2 contains the values for *four* of these functions at a series of r values and Figures 2.2 and 2.3 illustrate the graphs which are obtained by plotting these values. (Also shown in Figure 2.3 are the $3s$ and $3p$ radial functions.) The author strongly recommends that the reader recalculates some of the figures in Table 2.2 from the expressions in Table 2.1 so as to 'get to grips' with the functions. Indeed this procedure should be followed for all functions encountered in this chapter. A striking difference between $1s$, $2s$, $3s$ on the one hand and $2p$, $3p$, $3d$ on the other should be immediately obvious. The former have *positive* values when $r = 0$ whereas $2p$, $3p$,

Table 2.1 *Radial functions*, $R(r)$, *for hydrogen-like systems*

Normalizing constants*	Function† $R(r)$	Principal quantum number (n)	Angular momentum quantum number (l)	Symbol
$2Z^{3/2}$	e^{-Zr}	1	0	$1s$
$\dfrac{Z^{3/2}}{2\sqrt{2}}$	$(2 - Zr)\,e^{-Zr/2}$	2	0	$2s$
$\dfrac{Z^{3/2}}{2\sqrt{6}}$	$Zr\,e^{-Zr/2}$	2	1	$2p$
$\dfrac{2Z^{3/2}}{81\sqrt{3}}$	$(27 - 18Zr + 2Z^2r^2)\,e^{-Zr/3}$	3	0	$3s$
$\dfrac{4Z^{3/2}}{81\sqrt{6}}$	$(6Zr - Z^2r^2)\,e^{-Zr/3}$	3	1	$3p$
$\dfrac{4Z^{3/2}}{81\sqrt{30}}$	$Z^2r^2\,e^{-Zr/3}$	3	2	$3d$

* The general expression for the normalizing constant is $Z^{3/2}\sqrt{\left(\dfrac{4(n - l - 1)!}{n^4[(n + l)!]^3}\right)}$

† r is given in atomic 'Bohr radius' units (1 atomic unit = 0·529Å).

$3d$ are *zero* at $r = 0$. This stems from the fact that the latter contain the factor r^{n-l-1} which, being zero at $r = 0$, causes the whole radial function to be zero at this point.

Now the importance of the point $r = 0$ is that it is the *centre of the nucleus* and so we must conclude that certain radial functions have considerable magnitude right up to the centre of the nucleus whereas others do not. It should be noticed that those with the former behaviour are associated with the l number zero (i.e. s-functions).

All radial functions are independent of the angles θ and ϕ and will therefore, for a given r, always have the same values in any direction. Functions with this kind of behaviour are said to be

Table 2.2 *Values of some radial functions*, R(r), *for hydrogen*

$r_{(a.u.)}$	1s	2s	2p	3d
0	2·000	0·707	0	0
0·2	1·636	0·575	0·037	—
0·4	1·339	0·463	0·067	0·001
0·6	1·096	0·366	0·091	—
0·8	0·898	0·284	0·109	0·004
1·0	0·735	0·214	0·124	—
1·2	0·602	0·155	0·134	—
1·4	0·493	0·105	0·142	—
1·6	0·403	0·063	0·146	—
1·8	0·330	0·029	0·149	—
2·0	0·270	0	0·150	0·018
2·2	0·221	−0·023	0·149	—
2·4	0·181	−0·043	0·147	—
2·6	0·148	−0·058	—	—
2·8	0·121	−0·070	0·141	—
3·0	0·099	−0·079	—	0·029
3·2	0·081	−0·085	0·132	—
3·4	0·066	−0·090	—	—
3·6	—	−0·093	0·121	—
3·8	—	−0·095	—	—
4·0	0·036	−0·096	0·110	0·038
4·2	—	−0·095	—	—
4·6	—	−0·092	—	—
5·0	0·013	−0·087	0·084	0·042
5·4	—	−0·080	—	—
6·0	0·005	−0·070	0·061	0·044
7·0	0·002	−0·053	0·043	0·043
8·0	—	−0·039	0·030	0·040
10·0	—	—	—	0·032
12·0	—	—	—	0·024

spherically symmetrical. At this juncture it is worthwhile emphasizing that the hydrogen-like radial functions *describe the behaviour of the single electron* of these atoms. It must not be imagined that an electron is 'under the peaks' of these radial functions. The radial functions themselves are *graphs* and *do not have any physical sig-*

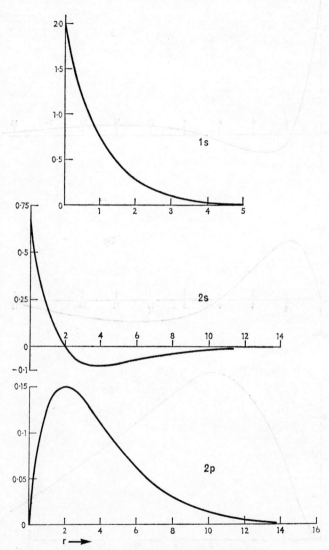

FIGURE 2.2. Radial functions for hydrogen

c

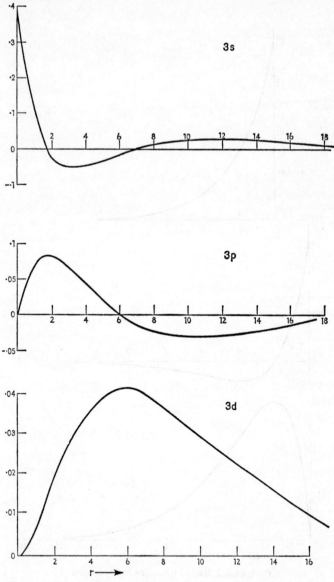

FIGURE 2.3. Radial functions for hydrogen

nificance. A more instructive way of regarding them will now be developed which will tie up with quantities which can be observed experimentally.

This discussion of functions is all very well but how do they relate to physical reality? After all, chemical compounds can be seen, touched and weighed. The answer to this problem is extremely elegant: although no apparatus can be set up to *measure* the wave function, ψ, of an electron in an atom or a molecule, experiments *can* be carried out which provide information about where the electrons are in the molecule. The *dipole moment* of a molecule, a measurable quantity, stems from electrical asymmetry in the compound, caused by uneven distribution of the electrons. We previously correlated the electron density at a point with the probability of finding the electron there. A second crucial correlation is that the *square of the wave function at a point is equal to the probability*, and hence equals the electron density at that point. This means that, if we refer back to Figures 2.2 and 2.3 and take any r value on these graphs, the electron density there will be the square of the value of the function at that point. It is common practice to equate the *square* of the wave function to the *electron density* in this way. This is an extremely valuable interpretation of ψ^2 and, although not *strictly* rigorous, it is one which will be frequently employed throughout this text. A more rigorous interpretation is the equation of ψ^2 to the probability. It is easy to see, quite generally, that the probability does correlate with the electron density because if there were 100% probability of finding an electron at point X then the density of electrons at X would be exactly one.

It is now interesting to recall that s radial functions had finite magnitude right up to the point $r = 0$ (the centre of the nucleus). That these functions do exist up to the nucleus means that electrons must permeate the nucleus and can interact with the nucleons. This is true for *all* s-type functions. At first sight it seems surprising that valence s-electrons, which in simple theories (e.g. Bohr) remain well outside the nucleus, are now described *inside* the nucleus. This is, however, only one of the interesting features of the new approach.

The complex-looking *normalizing constants* which appear before radial functions can now also be fitted into the theory. We argue thus: if we are trying to locate the electron we *must* find it if we look everywhere in the space available to it. This tells us that there is 100% probability of finding the electron *somewhere* in the space. Now, bearing in mind that probability = ψ^2, there is a convenient mathematical way of searching through a region: we integrate over it and the present 'search' is thus expressed as

$$\int_{\substack{\text{over all} \\ \text{space}}} \psi^2 \, d\tau = 1.$$

Here $d\tau$ is a small element of the space. The left-hand side expresses the search through all the space and the right-hand side says that this search must find the electron, i.e. the probability is 100%. The *normalizing constant* placed in front of the radial function ensures that this relationship is satisfied and unity is obtained when the integration is performed. It is a good idea to try this out on a function already plotted. The $1s$ radial function for hydrogen (in which $Z = 1$) has the form $\psi_{1s} = 2\,e^{-r}$, and here the number '2' is the normalizing constant. Let us first leave this number out and work out the value of the integral of the square of the function. The variable r runs from 0 to ∞ (Figure 2.1) and so the integral is

$$\int_{\substack{\text{over all} \\ \text{space}}} \psi^2 \, d\tau = \int_0^\infty [(e^{-r})(e^{-r})]r^2 \, dr = \int_0^\infty e^{-2r} r^2 \, dr$$

$$= \frac{2!}{2^3} = \frac{1}{4}$$

(it should be noticed that $d\tau = r^2 \, dr$ – not just dr). Using this function *without* its normalizing constant would therefore lead to our saying that the probability of finding the electron between $r = 0$ and $r = $ infinity is 25%. Obviously this answer is devoid of meaning because there isn't anywhere else to look! Multiplying the wave function by the number 2 takes account of this because $\psi^2 = 4\,e^{-2r}$, the integral of e^{-2r} is $\frac{1}{4}$ and so the overall answer is unity when the normalizing constant is put in. It should be remembered that this particular constant normalizes *only* the radial part of the total wave function. Obviously the *form* of the radial function is not changed by normalization but its *actual values* at different points will be.

The integral 'over all space' contains the element 'dτ' but the integration actually carried out was with respect to the radial variable r. This was done because we were testing out the normalization concept only on the radial function. Usually 'dτ' will be equivalent to a *volume* when the other two variables (here θ and ϕ) are added in. Now the volume expressed in Cartesian coordinates is 'length × breadth × height' and so

$$dv = dx \times dy \times dz.$$

The polar coordinates r, θ and ϕ are related to a volume expressed not as a 'square box', but as a sphere, i.e. $\frac{4}{3}\pi r^3$. There are, of course, other sets of coordinates which can express a three-dimensional situation and each set is associated with a 'volume form', e.g. yet another set of coordinates is related to an ellipsoid.

Because of the numerous ways of expressing volume it is usual to refer to its element quite generally in an integral as 'dτ'. It should further be noticed that, although $dv = dx \times dy \times dz$, the corresponding expression in polar coordinates is not simply $dr \times d\theta \times d\phi$. There must also be present certain *transformation factors* which convert Cartesian to polar coordinates.

Thus $$dv = dx\, dy\, dz = r^2\, dr \sin\theta\, d\theta\, d\phi.$$

This is a mathematical technicality and need not be pursued further here.

The Radial Distribution Function

There is another way in which the electron density–probability concept may be expressed. Instead of considering the square of the radial function which, as we saw, tells us about the probability of finding an electron at a *point* we often employ a new function called the *radial distribution function* (R.D.F.). This is defined as $4\pi r^2\, R(r)^2$, i.e. the square of the radial function multiplied by $4\pi r^2$. Values of the radial distribution functions derived from the 1s-, 2s-, 2p- and 3d-functions of hydrogen are given in Table 2.3 and are plotted in Figures 2.4 and 2.5 together with the 3s- and 3p-functions. The interpretation of these functions is subtly different from that of $R(r)^2$ alone. The R.D.F. contains the factor r^2 and so is *always* zero at $r = 0$ even for s-orbitals. This differs from $N(r)^2$ which for s-orbitals is not zero at $r = 0$. We interpret the radial distribution

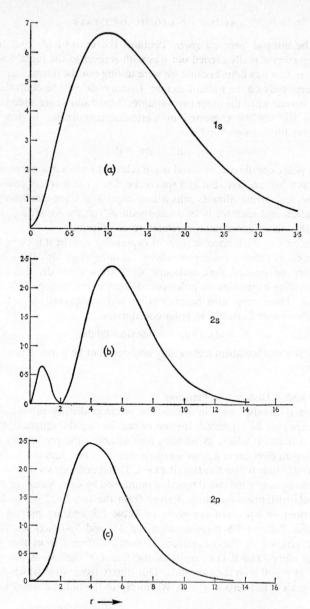

FIGURE 2.4. Radial distribution functions for hydrogen

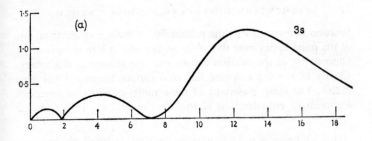

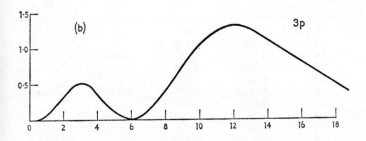

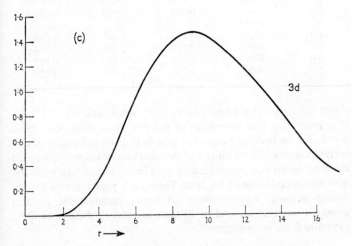

FIGURE 2.5. Radial distribution functions for hydrogen

function at a radius r as the probability of finding an electron, not at the point r, but over the *whole surface of a sphere of radius r* or alternatively as the electron density over the surface of this sphere. Clearly at $r = 0$ the sphere has zero surface because it has zero radius. The most important of these radial distribution functions for primary consideration is the $1s$ (Figure 2.4). This shows a

Table 2.3 *Values of radial distribution functions,* $4\pi r^2 R(r)^2$, *for hydrogen*

r(a.u.)	$1s$	$2s$	$2p$	$3d$
0	0	0	0	0
0·2	1·340	0·166	—	—
0·4	3·594	0·429	0·009	—
0·6	5·421	0·605	—	—
0·8	6·460	0·647	—	—
1·0	6·766	0·575	0·192	0·000
1·2	6·531	0·434	—	—
1·4	—	0·272	—	—
1·6	5·217	0·129	—	—
1·8	—	0·337	—	—
2·0	3·662	0	1·128	0·017
2·4	2·370	0·130	—	—
2·8	1·449	—	—	—
3·0	—	0·700	2·100	0·099
3·2	0·850	—	—	—
4·0	0·268	1·833	2·441	0·288
5·0	0·057	2·370	2·193	0·564
6·0	0·011	2·231	1·673	0·865
7·0	0·002	1·746	1·140	1·120

single maximum at a point where $r = 1$ atomic unit (i.e. 0·529 Å). It is interesting that the radius of the first Bohr orbit (i.e. where $n = 1$) for the hydrogen atom has *exactly* this value. Because of this correlation this unit becomes a convenient unit of length: it is called the *Bohr radius* and symbolized by a_0. The coincidence between the new results and those of the Bohr Theory is a great triumph for the former because, whereas Bohr arbitrarily assumed the integral quantum numbers n, l and m, in the wave theory these arise quite naturally from the mathematics.

There are other differences between the two theories which are worth noting. The chief of these lies in the comparative interpretation of orbits and orbitals. The $1s$ Bohr orbit was circular and the electron revolved around the nucleus in a fixed path of radius a_0. The new meaning of the Bohr radius is that it is the distance at which there is maximum probability of finding the electron over the surface of a sphere. Figure 2.4 (a) shows rather more than this though; there is also high probability (though not of course as great as at the Bohr radius) of finding the electron on either side of this radius. This is a consequence of the Heisenberg uncertainty principle because an exact momentum and energy can be calculated for an electron described by the $1s$ wave function so its position must be correspondingly indefinite. The electron does not then have a definite path or orbit at radius a_0 but may be found both inside and outside this radius. In terms of electron density it means that a $1s$ electron is, as it were, spread out over the whole of space but most is concentrated around the Bohr radius. When only one electron is involved, as here, the density interpretation is less appealing; it is more comprehensible to talk in terms of probability rather than to try to think of an electron as being 'smeared out' over the whole of space. The radial function is the one which governs the most probable physical distance of an electron from the nucleus and so determines the size of an atom. The meanings of the radial functions and their squares are of fundamental importance in the theory of atoms and molecules; they are not easy to grasp and should be studied from all aspects with great care.

Interpretation of *only* the single maximum of the $1s$ R.D.F. introduces a new difficulty because a glance at the other graphs in Figures 2.4 and 2.5 suffices to see that there is sometimes more than one 'hump' on the radial distribution function, albeit there is one bigger than the others. Moreover, the *shapes* of the peaks are not constant – sometimes the maxima are fairly sharp (Figure 2.4 (a)) and sometimes rather flat (Figure 2.5 $(b$ & $c)$). Because of these subsidiary features of the functions it seems that, to interpret the size and position of *only* the highest peak is neglecting other properties of the functions: indeed it might be suspected that the point r at which the highest maximum occurs is perhaps not a particularly fundamental point after all. Certainly it does not lead to anything particularly useful if it is compared from orbital to orbital. It would seem far better to *average* in some way the probability of finding the electron *over the whole range of the variable r*, and so find the *average* distance of the electron from the nucleus in any direction for a particular orbital. This quantity *can* then be compared from orbital to orbital because it would take into account all the apparently 'odd' features inherent in the wave functions. In order to carry out this averaging process we take what is called the '*quantum mechanical average value*' of *r*. This procedure need not be gone into further here; it will suffice to know that it is the rigorously correct way to interpret radial functions.

Effect of Nuclear Charge on the Radial Distribution Function

We now know that the radial distribution functions govern the average distance of the electron from the nucleus. Hence if an R.D.F. has a maximum near $r = 0$ the electron is close to the nucleus whereas if the former is spread out over a larger range of r the electron is correspondingly further away. It is therefore of great interest to study the effect on the R.D.F. of varying the charge (Z) on the nucleus. The nuclear charge enters into both the normalizing constant and the exponent of the function (Table 2.1). In Figure 2.6 curves A and B are the R.D.F's for an electron in a 1s-orbital with $Z = 1$ (H) and $Z = 2$ (He$^+$). It can be seen immediately that the increased nuclear charge of He$^+$ pulls in the 'hump' of the R.D.F. to 0·5 Bohr radii instead of 1 Bohr radius as in hydrogen. Physically speaking this means that the He$^+$ species is a smaller entity than is neutral hydrogen. Curve C is the R.D.F. for a species with $Z = 1·5$. It would represent roughly the 1s-orbital of the *neutral* helium atom which has two electrons. Here neither of the two electrons 'feels' the whole of the nuclear charge of +2 units but is partly screened from it by the other. (Notice also, on the other hand, one electron does not 'block-off' the *whole* of one unit of nuclear charge from the other.) This screening arises because both electrons can be described by the same type of wave function and have the same average distributions in space. The overall result is that both experience something like $Z \approx 1·5$. Screening is always present in atoms or ions which have more than one electron and it introduces complexities with which we cannot deal here. However, what has been shown is that the physical *size* of an atom or ion is governed by the R.D.F. of its 'outermost' electron which in turn varies with the apparent or *effective* nuclear charge determined by mutual screening.

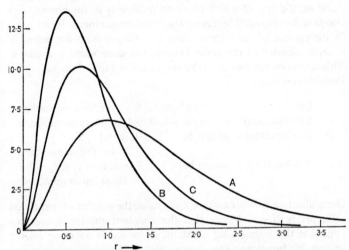

FIGURE 2.6. Effect of nuclear charge on the hydrogen-like R.D.F.

The Angular Dependence Function

The second part of the total wave function for an electron is called the angular function because only the angular variables θ and ϕ are involved in its make-up. This fact must be constantly borne in mind at all stages: angular functions are subject to more misinterpretation than any other concept in the theory. When there is any doubt as to the relation of the functions to physical reality then the answer will usually become clear by appealing to the mathematics – always simple because only sine and cosine functions are involved.

The composite angular function is made up of *three* separate factors and can be written in the form,

Angular function = constant × (a simple polynomial in cos θ or sin θ) × (sin $m\phi$ or cos $m\phi$) e.g. k × $\sin^2 \theta$ × cos 2ϕ or k' × sin θ × cos ϕ.

Firstly we will consider the multiplying constant. This is, of course, the now familiar normalizing constant for the angular function and it is necessary in order to maintain a sensible correlation between probability and electron density. It contains two integral numbers, l and m, the first of which occurred previously in the normalizing constant for the radial function. The quantum number m is new to us in the context of wave theory and it corresponds to the *magnetic quantum number* of the Bohr Theory: the same name is retained. This quantum number can take the values $-l$ to $+l$ (including 0). Possible cases are:

$l = 0$ (s-function) m can only equal 0   i.e. one m number
$l = 1$ (p-function) m can be $-1, 0, +1$ i.e. three m numbers
$l = 2$ (d-function) m can be $-2, -1, 0, +1 +2$
                                              i.e. five m numbers
$l = 3$ (f-function) m can be $-3, -2, -1, 0, +1, +2, +3$
                                              i.e. seven m numbers.

Generalization is now easy; for any l value the number of associated m values is $2l + 1$. Notice that the quantum number n does not appear in this normalizing constant, or indeed in anything related to angular functions. This means that a $1s$-angular function has precisely the same form as a $2s$-, a $3s$- or $4s$-. Similarly a $2p$-angular function is the same as a $3p$- or a $4p$-.

As with the radial functions there are a *set* of angular functions depending on (i) what *power* of $\sin \theta$ or $\cos \theta$ is involved and (ii) the value of m. We shall discuss these together. Firstly in an angular function corresponding to the quantum number l the power of $\sin \theta$ or $\cos \theta$ is l: Secondly the values of m are determined by the l number. Thus for:

$$l = 0, m = 0$$

the function contains $\cos^0 \theta$ or $\sin^0 \theta$ and $\cos 0\phi$, i.e. the whole is unity and therefore the angular function must be simply a constant, its overall value being determined by its normalizing constant. This is the s-angular function. This function is independent of the angles θ and ϕ and so, like a radial function, is spherically symmetrical.

$$l = 1, m = 0, \pm 1$$

There are $2l + 1 = 3$ angular functions in this set:

$$k \cos \theta \cos 0\phi \quad \text{i.e. } l = 1 \quad m = 0$$
$$\left.\begin{array}{l} k' \sin \theta \sin \phi \\ k'' \sin \theta \cos \phi \end{array}\right\} \quad l = 1 \quad m = \pm 1.$$

The first function simply reduces to $k \cos \theta$ and so does not depend on the variable ϕ at all. The last pair of functions are associated with *both* the m numbers ± 1. They are both *real* functions and so can be plotted. Now the expressions for x, y and z in polar coordinates (Figure 2.1) are $x = r \sin \theta \cos \phi$, $y = r \sin \theta \sin \phi$, $z = r \cos \theta$ so that, leaving aside the r variable in these expressions, the above angular functions *without* their normalizing constants are equal to x, y and z. Since these are p-type angular functions (they have $l = 1$) they are called the p_x-, p_y- and p_z-orbitals. Later on these will be plotted out.

$l = 2, m = 0, \pm 1, \pm 2$

There are five of these and they are the five d-angular dependence functions. Their naming derives from the equivalence of the expressions for them in polar coordinates with x, y and z. As an example consider the function

$$\sin \theta \cos \theta \cos \phi = \underbrace{(\sin \theta \cos \phi)}_{x} \times \underbrace{\cos \theta}_{z}.$$

This orbital is hence called the d_{xz}-orbital. Note that here the factor $\sin \theta \cos \theta$ derives from $l = 2$ just as if it were $\sin^2 \theta$ or $\cos^2 \theta$, i.e. it is a *quadratic* factor just as $x^2 - 2 = 0$ and $xy = 4$ are *both* quadratic equations. Another example is $\sin^2 \theta \cos 2\phi$ which is called the d_{xy}-orbital. Leaving out the normalizing constants the five d-angular functions are,

$$
\begin{array}{ll}
\sin^2 \theta \cos 2\phi & d_{xy} \\
\sin^2 \theta \sin 2\phi & d_{x^2-y^2} \\
\sin \theta \cos \theta \sin \phi & d_{yz} \\
\sin \theta \cos \theta \cos \phi & d_{xz} \\
3 \cos^2 \theta - 1 & d_{z^2}
\end{array}
$$

Two points are worth noting about these functions: (*i*) the subscripts always involve quadratic factors and (*ii*) the fifth function, like the p_z-angular function, has no dependence on the angle ϕ.

In order to understand the form and properties of angular functions in general it is a good idea to plot them out graphically. We shall plot only the p_z and d_{z^2} but the same principles will operate for any other.

Graphical Plots of Angular Functions

We propose to plot the p_z-function because it is the simplest (as it does not involve the angle ϕ). On first encounter with $\cos \theta$ in school it is usual to plot it as a *wave* in which the *angle* θ is represented as a *linear* distance along the abscissa (Figure 2.7). There is a second,

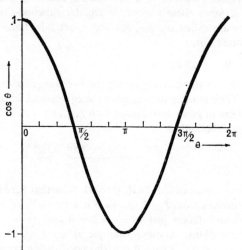

FIGURE 2.7. Linear ('wave') plot of $\cos \theta$

more interesting, way this can be done which is known as a *polar plot*. First the z- and y- (or x-) axes are drawn up as in Figure 2.8. The polar angle θ (Figure 2.1) is the angle between the z-axis and a line OP. Starting at $\theta = 0$, $\cos \theta$ is calculated and multiplied by its normalizing constant N $\{= \sqrt{6}/2\sqrt{(2\pi)}\}$. Table 2.6 lists the values of p_z for such a set of angles θ. The procedure for plotting the function is to measure out along OP the appropriate number in the table ($\cos \theta$) for the angle θ which OP makes with the z-axis. The negative

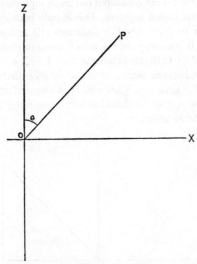

FIGURE 2.8. Definition of axes and angle θ

Table 2.4 *The angular functions p_z and d_{z^2}*

θ (deg)	p_z	d_{z^2}
0	0·489	0·631
20	0·459	0·520
40	0·374	0·240
60	0·244	−0·079
80	0·085	−0·287
90	0	−0·315
100	−0·085	−0·287
120	−0·244	−0·079
140	−0·374	0·240
160	−0·459	0·520
180	−0·489	0·631

numbers in Table 2.4 are measured out as though they were positive along OP but are called negative. This is only because there are no negative rulers to draw negative distances (!!) and need cause no alarm because it is exactly the same as is done in plotting the graph $y = x$ (Figure 2.9). In the latter when $x = -1, -2, y = -1, -2$, but these y and x values are measured out as *positive* distances along the axes but *called* $-x$, or $-y$. Thus y has positive and negative regions just as does the p_z-angular function albeit the latter is plotted in a way which appears different.

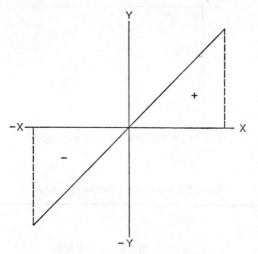

FIGURE 2.9. Graph of $y = x$

Figure 2.10 depicts the curve generated by taking angles θ between $\theta = 0$ and $\theta = 180°$ (i.e. radians) and is a polar plot. Finally because ϕ does not enter into the expression for p_z at all this plot could be obtained for *any* angle ϕ. The full function is thus generated by rotating the figure by $360°$ around the z-axis (it will be recalled that the angle ϕ is defined around this axis).

There are a number of features in this complete diagram which merit discussion:

(*i*) It is a *three-dimensional* graph consisting of *twin spheres in tangential contact at the origin.* The upper sphere represents the positive part of the function in the range $\theta = 0°$ to $\theta = 90°$ (i.e. $0 - \pi/2$ radians) while the lower lobe is the negative part of the function between $\theta = 90°$ and $\theta = 180°$ ($\pi/2 - \pi$). In both parts ϕ runs from 0 to 360°.

(*ii*) The greatest distance from the origin to the surface of the graph is directly along the *z*-axis.

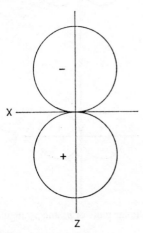

FIGURE 2.10. Normalized polar plot of $\cos \theta$. (p_z angular function)

(*iii*) If functions p_x and p_y were plotted we would have to take a series of values of θ *and* ϕ and would then derive three-dimensional graphs identical in form to the one we have plotted. They would differ, however, in that they would be directed along the *x*- and *y*-axes respectively and so would have their maximum extensions in these directions (Figure 2.11). Thus these three functions p_x, p_y, p_z, are equivalent to each other in every way except direction and, because of this, each separately corresponds to the same total energy, E, in the Schrödinger equation. Such a set of functions are said to be a *degenerate* set and in this case, as there are three *p*-functions, they constitute a *triply degenerate set*.

Precisely the same sort of procedure may be followed to construct diagrams of *d*-angular functions. The simplest example of these is

D

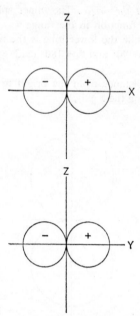

FIGURE 2.11. Polar plots of p_x and p_y angular functions

d_{z^2} because, like p_z, it has no dependence on the angle ϕ, its functional form is $N(3 \cos^2 \theta - 1)$. All that is required is to take a set of angles θ from 0 to 180°, find the cosine of each, square it, multiply it by 3, subtract one from the result and finally multiply the whole by the normalizing constant N ($= 10/4$). The set of numbers is plotted as before for the p_z-angular function. Table 2.4 gives figures for various angles θ between 0 and 180° and Figure 2.12 illustrates the full polar plot of d_{z^2}. (As before, the whole three-dimensional function is generated by rotating this diagram around the z-axis by 360°.)

The d_{z^2}-function possesses a number of interesting features. Firstly it is similar to p_z in that it has its maximum extension along the z-axis. However, the *sign* of this function in different regions does not follow the same pattern as p_z; it is positive along *both* $+z$ and $-z$

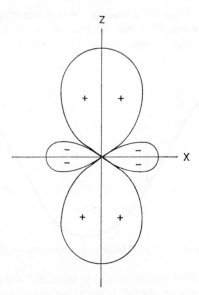

FIGURE 2.12. Normalized polar plot of $(3 \cos^2 \theta - 1)$. (d_{z^2} angular function)

(because it contains z^2) but has a negative 'collar' in the region of the xy plane. Bear in mind that these signs are only those of the function, i.e. the graph, and have nothing at all to do with the charge of the electron. Thus the d_{z^2}-angular function between 0 and 180° could have been plotted in the old 'wave' way (Figure 2.13). Regions A and C correspond to the two positive lobes of Figure 2.13 and region B to the negative 'collar'. Notice that the function crosses the abscissa near 55°; this is the angle θ' in Figure 2.13.

If the other four d-functions, associated with the m numbers ± 1, ± 2, were drawn out for various values of θ *and* ϕ the set shown in Figure 2.14 would be produced. These four all have similar shapes to each other and differ only in their relative orientation: the d_{xy}, d_{xz} and d_{yz} are equivalent and lie between the axes in the xy, xz and yz planes respectively. The $d_{x^2-y^2}$-function has the same shape as these but is directed *along* the x- and y-axes. These functions all exhibit four lobes but it must be stressed that they are each still *one single function*. This may be checked in the following way: consider d_{xz}, which has the functional form $\sin\theta\cos\theta\cos\phi$: if ϕ is

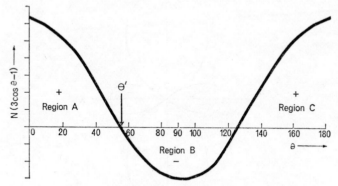

FIGURE 2.13. Linear plot of $(3\cos^2\theta - 1)$

put equal to 0 the function reduces to $\sin\theta\cos\theta$ and a silhouette in the xz plane may be plotted. If the 'wave' form is now drawn for

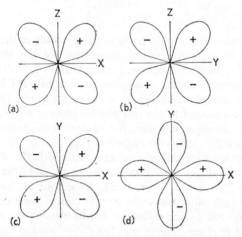

FIGURE 2.14. The d_{xz}, d_{yz}, d_{xy}, and $d_{x^2-y^2}$ angular functions

$\theta = 0$ to $\theta = 180$ (Figure 2.15), the function is positive between 0 and 90° (Region (A)) because both $\sin \theta$ and $\cos \theta$ are positive in this range and their product is a maximum at 45°; it becomes negative from 90° to 180° (Region (B)) because $\cos \theta$ is negative in this range. This coincides exactly with the polar plot of d_{xz} in Figure 2.14. The fact that the last four d-functions are not shaped like d_{z^2} does not mean that they are fundamentally different. All five d-functions are

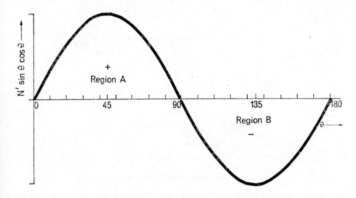

+
Region A

Region B
−

FIGURE 2.15. Linear plot of $\sin \theta \cos \theta$

equivalent just as were the three p-functions. They therefore form a *five-fold* degenerate set.

It may seem strange to leave a consideration of the s-angular function till last but in some ways it is the most difficult to grasp. The s-angular function does not contain the variables θ or ϕ *at all* and is only a simple positive constant $(1/2\sqrt{\pi})$. A polar plot shows that it has this same positive value *everywhere independently of direction*; this must be true as it does not contain θ and ϕ at all. The polar representation is thus a *sphere* of radius $1/2\sqrt{\pi}$ (Figure 2.16 (*a*)). A 'wave' plot of an *s-angular function* is particularly simple; in the xz plane (where $\phi = 0$) it has one value only for all angles θ and so is just a straight line (Figure 2.16 (*b*)). We shall find it helpful to refer to these wave plots again because polar plots of angular functions are frequently assigned mystical properties which they do not possess and reference to an ordinary 'wave' plot helps to bring the subject back to earth. A point of difficulty which arises with the *s*-angular function is that it is often confused with radial functions

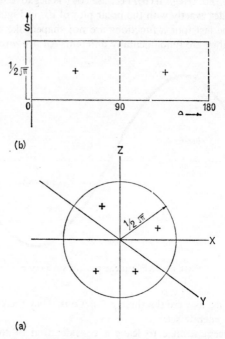

(b)

(a)

FIGURE 2.16. Polar and linear plots of the *s*-angular function

which are *all* spherically symmetrical and can all be roughly drawn rather like the *s*-angular function. Much care must be exercised to avoid confusion of the two and it must always be remembered that an atomic orbital is the *product* of two functions, the radial and the angular.

Misconceptions of Angular Functions

We continue this chapter now with a discussion of some misconceptions regarding angular functions. These are well illustrated by taking as example the p_z-angular function (Figure 2.10). This is often depicted as a 'figure 8' or 'dumb-bell' (Figure 2.17) having an upper positive and a lower negative lobe and it is implied (or stated) that an electron is inside this figure. Furthermore, it is suggested that distances in this diagram have physical meaning and can in some way be correlated with distances between atoms or from nuclei. These misconceptions may readily be corrected by invoking no more than very elementary mathematics. Let us take them point by point.

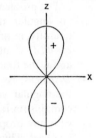

FIGURE 2.17. 'Dumb-bell' representation of a p-angular function

(1) It must be recognized that the diagram drawn represents only the *angular* part of the wave function. Thus the polar diagram of a p-function is a function of the polar angle θ and the azimuthal angle ϕ and does not contain the radial variable r. Hence, it may not be related to *any property which involves physical distances*.

(2) *'Figure 8' concept:* It is readily shown that a p-function does not have this shape because p_z has the functional form $\cos \theta$ (no $\sin \phi$ or $\cos \phi$ factor) and, as we have seen, $\cos \theta$ may be plotted in two distinct ways one of which is the polar diagram. Clearly then, the solid function cannot possibly be a 'figure-8-of-revolution'. The whole concept would present no problem if, on first encounter in school, $\cos \theta$ was represented by a polar plot as above rather than as a continuous wave.

(3) *Electrons 'inside' the orbital:* It is common practice to talk of electrons being 'inside' orbitals. This is harmless if it is realized that what is meant is that their energies and movement are described or fixed by a wave function or probability distribution. Unfortunately, the shape of the p-angular function (for example) lends plausibility to the *physical* idea of an electron moving inside one or both of the lobes. This tries to give meaning to distances measured from the origin to points inside the lobe and also raises irrelevant questions like 'if the

electron has zero chance of being at the origin, then how does it get from one lobe to the other?'

The problem is immediately resolved by realizing that such an angular dependence function is only a mathematical quantity, in one case $\cos \theta$, and so how can an electron be in $\cos \theta$? It would seem ludicrous to make the same suggestion about the function plotted as a wave.

(4) *Distances:* As stated, the *p*-function plotted out involves only the angular variable θ and is an angular wave function. It contains no radial coordinate, r, and so cannot be correlated with any property of the atom in which physical distance is involved. Thus distances on the polar diagram bear no relation to distances from the nucleus in an atom. Realization of this fact enables another misconception to be corrected, viz.

The impression that the surfaces of the two spheres should be 'fuzzy' and indefinite and that they only represent boundaries inside which a 'given percentage of electronic charge exists'.

This has arisen from a confusion of the angular and radial functions. The former only are under discussion and we have already proved that the function has a definite boundary surface and cannot possibly be 'fuzzy'. We should indeed be surprised to learn that we were never told the truth about a sine or cosine wave plot and that it should really be drawn as a vague and not a definite line!

The radial function, on the other hand, has spherical symmetry for all orbitals and a decreasing exponential term in its make-up. Hence it is asymptotic to the radial distance axis. Whether or not one considers this function as a sphere within which a fraction of the electronic charge exists is a personal matter, though in the author's view this adds nothing to its understanding and, indeed, tends to lead to the confusion with which this section is concerned.

How then do we interpret the polar diagram? Firstly, it is clear that distances on the diagram from the origin to points inside the lobes have no meaning whatsoever because of the way the function was plotted. The absurdity of interpreting such points inside the function is readily seen by referring back to Figure 2.7. The only way in which a wave function can be correlated with physical reality is to consider its square: this is proportional to the probability of finding an electron in a certain direction, or at a given distance, if the function is a radial function. This probability may reasonably (though not rigorously) be equated to the electron density in that direction or at that distance. Thus, the true interpretation of an angular function is that the square of the length on the polar diagram from the origin to any point on the surface of the function is the total probability of finding the electron at some radial distance between zero and infinity in that direction.

It may help in understanding the angular functions if one imagines the following (hypothetical!) situation. Suppose that one was standing on the atomic nucleus and was asked to point in the most probable direction in which an s electron would be found. Any direction would be appropriate because the probability of finding such an electron is equal in all directions. However, if it was a 'p_z electron' that was being referred to then the 'best bet' would be to point along the positive or negative z-axis because the p_z-function has its maximum extension (and hence its maximum square) along the z-axes.

The final topic for consideration concerns the *nodes* which arise in the radial and angular functions. A node in a function is a point where the function changes sign, positive to negative or vice versa. The number of nodes in the radial functions may be counted from Figures 2.2 and 2.3. Thus,

$$1s \; (n = 1, l = 0) \quad \text{no nodes}$$
$$2s \; (n = 2, l = 0) \quad \text{1 node}$$
$$2p \; (n = 2, l = 1) \quad \text{no nodes}$$
$$3s \; (n = 3, l = 0) \quad \text{2 nodes}$$
$$3p \; (n = 3, l = 1) \quad \text{1 node}$$
$$3d \; (n = 3, l = 2) \quad \text{no nodes}$$

In general the number of nodes in a radial function is $n - l - 1$. For the angular functions

s-functions are everywhere positive and so have *no* nodes (Figure 2.16), *p*-functions have one positive and one negative lobe and therefore have *one* node (Figure 2.10),

d-functions change sign twice and hence have *two* nodes (Figure 2.13).

There are thus *l* nodes in an angular function. Adding these two results together, the total number of nodes in the whole wave function is $(n - l - 1) + l = n - 1$. Thus for hydrogen-like wave functions *the total number of nodes depends only on n and not on l.*

It is very important to understand the significance of the nodes in a wave function. The total energy of an electron in a hydrogen-like atom depends on the wave function describing it and the energies associated with wave functions rise in ascending order of the number of nodes. Thus the 1*s*-function, with no nodes, has the lowest energy of all. The electronic energies for a hydrogen-like system increase with *n* (see next section). The information provided by a count of the number of nodes in a wave function is extremely useful and will be applied widely in a later study of the energy levels of molecules (Chapter 5).

Electronic Energy Levels for Hydrogen-like Systems

Apart from a brief reference the *energies* of the orbitals have not yet been discussed quantitatively. The total energy of an electron (E) appears in the Schrödinger equation and one of the incentives for solving the equation was that, having done this, we would be able to calculate the total energy of an electron described by a particular wave function. It turns out, for hydrogen-like atoms, that a set of values of E exist and are given by the equation,

$$E_n = \frac{-2\pi^2 m e^4 Z^2}{h^2 n^2}.$$

For hydrogen itself $Z = 1$ and in the 1*s* level, $n = 1$ hence

$$E_{H_{1s}} = \frac{-2\pi^2 m e^4}{h^2 \times 1}.$$

It follows that the energy of the *n*th orbital of hydrogen is $E_{H_{1s}} \times 1/n^2$, i.e. it depends only on *n*. No influence on the energies is exerted by *l* and the energies of, say, the 3*s*, 3*p* and 3*d* levels are identical. (This result does not hold when there is more than one electron in the atom.) The degeneracy of the triple and five-fold sets of p- and d-orbitals in hydrogen is confirmed by this result. Situations arise (in chemical compounds) where the degeneracy of these sets is altered so that they no longer all have the same energy and when this happens the degeneracy is said to have been *lifted*.

The numerical value of $E_{\text{H}1s}$, $-13 \cdot 6$ eV, is the total energy of an electron in the $1s$-orbital of a free hydrogen atom. It is important to realize that the hydrogen atom possesses a full range of higher atomic orbitals but these are not normally occupied. If the single electron were in the $2p$-orbital its energy would be $-13 \cdot 6/n^2$ $= -13 \cdot 6/4 = -3 \cdot 4$ eV. That the total energies are negative means that the hydrogen atom is more stable than the separated proton and electron.

The stabilizing energies of electrons in the orbitals of hydrogen-like atoms which have only one electron but more than one proton (He^+, Li^{++}) are *greater* than for the same orbital of neutral hydrogen. The reason for this is that the factor Z^2 appears in the numerator of the energy equation and, e.g. the energy of the $1s$-orbital of He^+ is $4 \times (-13 \cdot 6)$ eV whilst that of its $2p$ level is $4 \times (-13 \cdot 6)/4$ $= -13 \cdot 6$ eV.

The dependence of the energy of these systems on n alone correlates with the total number of nodes in the wave functions and as the number of nodes increases so the electronic binding energy diminishes.

Quantum Numbers for Atoms

It is worthwhile to collect together and summarize the significance of the quantum numbers introduced in this chapter. These were n the principal quantum number, l the angular momentum quantum number, and m the magnetic quantum number. There is also a fourth quantum number relevant to the behaviour of electrons, i.e. the spin quantum number. It did not emerge from the treatment because the form of wave equation initially considered contained only the space coordinates of the electron not its spin coordinates. The spin quantum number takes only the values $\pm\frac{1}{2}$ and these correspond to the electron's having two directions of spin. It is usual to designate these as α or β spin.

Principal quantum number n – in the hydrogen atom it governs the energy of the electron in any orbital. It also determines the form of the radial function and hence the average distance of the electron from the nucleus in any orbital it occupies.

Angular momentum quantum number l – governs the angular momentum of an electron in an orbital and the form of the angular function. It should be noted that the number does not now relate to the ellipticity of the orbit as in Bohr Theory. Combination of the n and l numbers enables creation of a system of notation for orbitals. These two numbers also determine the normalizing constants for radial functions.

Magnetic quantum number m – this is related to l and determines the angular momentum component in a given direction. Since it governs a component only it cannot be greater than the *total* angular momentum number and it therefore is limited to a set of values running from $-l$ to $+l$.

The numbers l and m make up the normalizing constants for the angular functions.

The Periodic Classification of the Elements

The Pauli principle states that no two electrons in a given atom can have the same set of numerical values for the quantum numbers. In terms of the new theory this means that two electrons can only be described by the same space wave function if one has α and one β spin.

The energy sequence for hydrogen-like orbitals depends only on the principal quantum number and the nuclear charge. For heavier elements, where there are many electrons, both these quantities require modification for the following reasons:

(*i*) Because each radial distribution function, independently of n, extends up to the nucleus, the electrons are 'mixed up' and so the well-defined n numbers of hydrogen become merged. The energy of an electron in a many-electron atom is inversely proportional not to n^2, but to an approximate or *'effective' principal quantum number* n^{*2}.

(*ii*) As was pointed out for helium (Figure 2.6), mutual screening of one electron by another reduces the nuclear charge felt by each (though not by one unit). The energy of an electron is not then proportional to Z^2 but to the *effective nuclear charge*, Z^{*2}. Moreover, the screening which a particular electron provides will differ depending on whether it occupies an s-, p- or d-orbital; so the energy of an electron in the atom now varies with l as well as n and with the extent to which other orbitals of the atom are occupied. The degeneracy of the $3s$, $3p$, $3d$ levels is thereby removed (Figure 2.18). The sequence of increasing orbital energies, apart from hydrogen, for all the elements up to atomic number $Z \approx 20$ is that of increasing $n + l$ rather than n alone (Figure 2.19). Coupled with

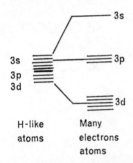

FIGURE 2.18. Lifting of degeneracy of hydrogen orbitals

the Pauli principle this rule enables us to construct the electronic arrangements of these elements. Overall neutrality of the atom requires that there must be the same number of electrons as protons. To arrive at the *ground state* electronic structure of the atom with atomic number Z we feed Z electrons two by two into the orbital set starting with the one of lowest energy. When the electronic arrangement violates this simple rule the atom is said to be *excited*.

The single electron of hydrogen is placed in the lowest level – the $1s$ atomic orbital. Helium ($Z = 2$) has two electrons and the second also goes into the $1s$-orbital. We write this as He $1s^2$. An electronic arrangement such as this is called a *configuration*. The $1s$-orbital can only take one α and one β electron and so is now filled. The next two elements, Li and Be, have electrons in the $2s$-orbital which is also filled at beryllium. Boron has five electrons and the 'extra' one is placed in the $2p$-orbital giving the electronic configuration $1s^2 2s^2 2p^1$. It is now straightforward to generate the structures of the next five elements by simply filling up the $2p$ level: thus carbon is $1s^22s^22p^2$, N $2p^3$, O $2p^4$, F $2p^5$, and Ne $1s^22s^22p^6$. When all the orbitals of a set are completely filled (e.g. He, Be, Ne) there is said to be a closed shell. The closed shell configuration of Ne is consistent with its chemical inertness. The next orbital in the energy scale is $3s$

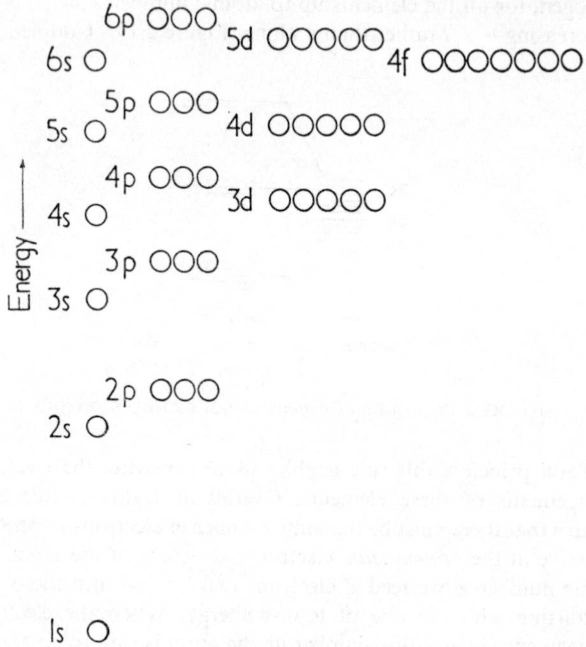

FIGURE 2.19. Energy sequence for the orbitals of many electron atoms

and sodium, the next element, has the electronic configuration $1s^22s^22p^63s^1$. The similarity of this structure (one unpaired s electron) to that of lithium is evident and leads to the chemical similarity of these two elements. The same pattern of configurations ensues (from which derive the main Periodic groups) until the next inert gas (argon $1s^22s^22p^63s^23p^6$) is reached.

Now in the hydrogen atom the $3d$ has the same energy as the $3s$

and $3p$ levels but in many electron atoms the $n + l$ sequence brings the $4s$ level lower than the $3d$. The next element in the series, potassium, therefore, has its outermost electron in the $4s$ level, and calcium ($Z = 20$) has this level filled. The $3d$-orbitals can now be filled sequentially and the first, scandium, has the configuration Ar-core $4s^2 3d^1$. The $3d$-orbitals proceed to fill up over the next nine elements and zinc, the last, has configuration Ar-core $4s^2 3d^{10}$. The $4p$ levels are next in the energy order with gallium having Ar $4s^2 3d^{10} 4p^1$ and krypton Ar $4s^2 3d^{10} 4p^6$, i.e. ending the period with all closed shells.

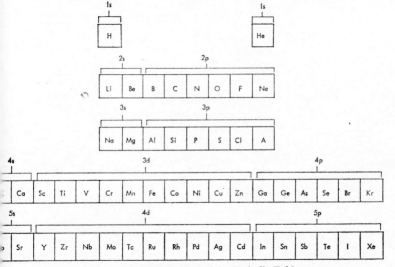

FIGURE 2.20. Early section of the Periodic Table

The same interchange of nd and $(n + 1)s$ now again operates and another set of elements with varying numbers of d electrons are found. This *whole block* of elements with incomplete d electron shells are called the *transition elements*. As before, at the end of each period is an inert gas having all shells closed. Figure 2.20 illustrates the build-up of the elements we have been discussing. After xenon a further factor intervenes; the $4f$- and $5d$-orbitals interchange in energy and the $4f$-orbitals become available. This gives rise to a

further sub-series of fourteen elements, the lanthanides. Consideration of the electronic structures of these elements is, however,
beyond the scope of the present text.

PROBLEM FOR CHAPTER 2

1. Construct a plot of the linear combination $(\sqrt{3}/\sqrt{2})$ $(\sin \theta$
   $+ \cos \theta)$.
   How is it related to the p_z- and p_x-angular functions?

CHAPTER 3

Valence-Bond Theory

We have now developed sufficient background to enable a study of the structures and bonding in molecules to be begun. In this context we will first examine closely the electron-pair concept of bonding in order to try and find out why electron pairing should bring about the bonding of two neutral atoms. Indeed, the opposite might have been anticipated since electrons, by virtue of their negative charge, repel one another and so pairing should lead to instability. Having rationalized this problem a more sophisticated description of co-ordinate, covalent and ionic bonds can be obtained within the framework of orbital theory.

There are two distinct ways in which one can approach the bonding problem: these have been called *valence-bond theory* and *molecular orbital theory* and the two differ quite markedly in concept. The treatment of molecular orbital theory is deferred until the next chapter; in this we shall investigate valence-bond theory. It is most convenient to consider diatomic molecules first of all. These are simple, they illustrate most of the principles involved, and are exemplary for solving the basic problems. We can then fairly easily extend the knowledge gained to larger molecules.

What then is the valence-bond approach? It is essentially dynamic in that initially the atoms making up the molecule are considered to be isolated at large distances from each other. They are then allowed to approach each other and gradually to interact as the charged bodies become closer and the electrons paired. With decreasing internuclear distance this interaction leads to a situation of lower energy than the starting one and this is why a molecule forms. However, in answering the primary question of why molecules are formed at all, the problem has only been pushed one stage further back – we still have to explain why electron pairing should bring about a lowering in total energy of the system.

E

Let us study a simple case first of all –the hydrogen molecule. The two hydrogen atoms (nuclie A and B, electrons (1) and (2)) are considered initially to be separated by a large distance such that they are effectively non-interacting. Suppose now we let them approach mutually. How will they interact with each other? Bearing in mind that the hydrogen atom is composed of a positively charged proton and a negatively charged electron then, as the atoms come

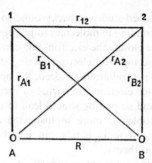

FIGURE 3.1. The hydrogen molecule

into close proximity (Figure 3.1), a number of effects will become strongly operative:

(i) electron (1) will repel electron (2)

(ii) nucleus A will repel nucleus B

(iii) nucleus A will attract *both* electrons (1) and (2)

(iv) nucleus B will attract *both* electrons (1) and (2).

All these effects can be expressed in an equation,

$$\varepsilon = -E_{A-1} - E_{B-2} - E_{A-2} - E_{B-1} + E_{1-2} + E_{A-B} \qquad (3.1)$$

where 'E_{A-1}', for example, represents the attraction energy between nucleus A and electron (1). It is usual to give negative signs to attraction energies and positive signs to repulsion terms. The way in which these terms act should be fairly obvious; the first two are attractive terms but *do not serve to hold the two atoms together*. This is because these two attractions are present even when the two atoms are separated so do not represent a gain in energy for the molecular situation. The second pair of terms are, however, new attractive terms and are stabilizing influences, helping to hold the two atoms together as a molecule. What is interesting is that, whereas in the isolated atoms electron (1) 'belongs' and is attracted only to nucleus A and (2) only to nucleus B, they are both now attracted to *both* nuclei. The final two terms in the equation are repulsion terms and will act to keep the two hydrogen atoms apart. They will favour the system's remaining as two isolated hydrogen atoms instead of forming a molecule.

These terms are not the only ones contributing to the molecular energy, i.e. 'ε' is not the *total* energy of the system. In addition the *kinetic energies* of the two electrons must be accounted for. However,

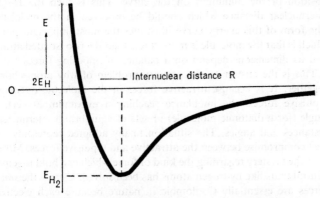

FIGURE 3.2. Energy curve for the hydrogen molecule

to a first approximation, these kinetic terms do not affect the reasoning and so they will not be explicitly brought into the following discussion. All the terms contribute to the overall energy of the hydrogen molecule and whether the system is stable or not will depend on their *relative* magnitudes.

Now as the two hydrogen atoms are allowed to approach each other the 'two-centre' terms (i.e. the last four in the energy equation) start to increase in effect but they do so at different rates. The most stable state of a system is that of lowest energy and so if the overall energy of the incipient molecule, taking into account all the interactions, is *less* than the initial energy of the hydrogen atoms then the new situation must be one of greater stability. Figure 3.2 shows the change in energy of the system when the internuclear distance, R, is decreased. The zero of energy on the diagram is that of the two separated hydrogen atoms. As the two atoms approach each other the energy of the system falls below the zero which means that the attractive forces between the atoms exceed the repulsive forces and so, overall, the system is stabilized. The same trend continues to a point of minimum energy but beyond this the repulsion between the two nuclei and between the two electrons dominates the situation and the molecule rapidly loses stability. Obviously what will happen in practice is that the system will acquire most stability by mutual approach of the hydrogen atoms as far as the position of the minimum on the curve. This is then the H—H internuclear distance which should be observed in the molecule. The form of this energy curve illustrates the most important point which is that the molecule is more stable than the two isolated atoms and its dimensions depend on a balance of opposing forces.

This is the answer to the primary problem of why atoms bond together. It is because attractive forces in the molecule outweigh repulsive forces, the imbalance reaching a maximum at certain single (for a diatomic molecule) or sets of equilibrium internuclear distances and angles. The situation finally adopted represents the best compromise between the attractive and repulsive forces. Moreover the mystery regarding the kind of force which can hold together neutral units like hydrogen atoms has been solved. Clearly the *static* forces are essentially Coulombic in nature because each electron is bound electrostatically to *both* nuclei. Now this is exactly the same type of bonding as was suggested for ionic compounds such as sodium chloride and so *there is no essential qualitative difference whatsoever* in the two cases. It is now straightforward to understand why some bonds are partially ionic – in these the pair of electrons bonding the two atoms (as above) is more located in one atom than the other, i.e. the electron density in the bond is unevenly distributed thus rendering the bond heteropolar. Further discussion of this point will be deferred until later when a more quantitative account can be presented.

The important point now revealed is that ionic and covalent bonds incorporate identical bonding forces, the 'pure' ionic and 'pure' covalent bonds simply being extreme examples of one general type of link. There is thus no magical quality in electron pairing. The analysis of bonding just presented is however, as might be suspected, too simple because all the interatomic forces have been discussed as though they could be described by purely classical electrostatics. This does not detract from the main tenets of the argument but to be more rigorous these forces must be assessed in a way which is in keeping with the theory of Chapter 2.

A More Quantitative Treatment of Hydrogen

Now that the major problem is settled a number of secondary ones arise; perhaps the most pressing of these are the shapes of polyatomic molecules and the heteropolarity of bonds. Before tackling these we will develop the terms in the energy equation rather more quantitatively. The nucleus–nucleus repulsion is the easiest to start with because, under the initial approximations, the nucleus is a single point charge of magnitude Ze with no wave properties. This is not strictly true (the de Broglie equation predicted that it should have associated wave motion) but, as its mass is ~2000 times that of an electron, the associated wavelengths are much smaller and neglect of its wave properties will cause no appreciable error. Thus the classical expression $(Ze \times Ze)/R$ which describes the repulsion energy of two positive charges of Ze units separated by a distance R is appropriate (for hydrogen this comes down simply to e^2/R).

The attraction energy of electron (1) for nucleus A is likewise obviously connected with $1/r_{A1}$ where r_{A1} is the distance between electron (1) and nucleus A (Figure 3.1). However, the position of electron (1) is not fixed; it can move anywhere in space and moreover it is not a point charge but must be described by a wave function. The correct way to represent the attraction of electron (1) for nucleus A is thus $\psi_A(1)^2/r_{A1}$ where $\psi_A(1)$ is the $1s$ wave function of electron (1) based on nucleus A and its square represents the electron density at any point. Now no point has been specified at which this electron density is to be measured, i.e. no numerical value has been assigned to r_{A1} – indeed this cannot be done since the electron is free to move over the whole of space. Thus the attraction energy for *every possible position* of the electron in space must be estimated and all the contributions added up to give a total answer. This would seem to be quite a job because r_{A1} can have an infinite number of values starting at 0 and going to infinity! There is luckily a way of doing all of these computations in one fell swoop; what we do is to integrate the expression over the variable r_{A1} from 0 to infinity *in all directions* around the nucleus A. This takes care of the addition of all the contributions.

Thus, the attraction energy between nucleus A and electron (1) is

$$\text{A.E.} = 4\pi \int_0^\infty \frac{\psi_A(1)^2 r_{A1}^2}{r_{A1}} \, dr_{A1}.$$

dr_{A1} is the small element of r_{A1} (Chapter 2). The factor 4π comes from considering all directions – it is not particularly important to us here, and the integration itself is only a technicality. An alternative way of writing this integral, which we shall use later, is

$$\text{A.E.} = \int_{\substack{\text{over all} \\ \text{space}}} \frac{\psi_A(1)^2}{r_{A1}} d\tau_1.$$

By similar reasoning the attraction energy between electron (1) and nucleus B is

$$4\pi \int_0^\infty \frac{\psi_A(1)^2 r_{B1}^2}{r_{B1}} \, dr_{B1}$$

and it follows that the terms

$$4\pi \int_0^\infty \frac{\psi_B(2)^2 r_{B2}^2}{r_{B2}} \, dr_{B2} \quad \text{and} \quad 4\pi \int_0^\infty \frac{\psi_B(2)^2 r_{A2}^2}{r_{A2}} \, dr_{A2}$$

will also arise. These are all attraction terms.

The final term representing the mutual *repulsion* of the two electrons is also couched in terms of electron densities. The distance r_{12} separates the electrons and the repulsion energy is expressed as the integral,

$$\iint_{\substack{\text{over all} \\ \text{space}}} \frac{\psi_A(1)^2 \, \psi_B(2)^2}{r_{12}} \, d\tau_1 \, d\tau_2.$$

Notice that this has to be integrated twice because it contains terms referring to both electron (1) and electron (2). This integral is, in

fact, rather difficult to evaluate and we shall not attempt ot carry
this out.

Wave Functions for the Hydrogen Molecule

The wave functions we used above ($\psi_A(1)$ or $\psi_B(2)$) were those
appropriate to *atomic* hydrogen. When dealing with a hydrogen
molecule rather than an atom should we expect the same atomic
wave functions to be just as appropriate? Clearly the wave functions
of the *isolated* atoms are separate hydrogen $1s$-functions but when
the atoms are close and 'interacting' we should really try to find a
way of writing *one* wave function which describes the *combined*
system. The question is, what is this wave function? Our first
reaction might be to try and solve the Schrödinger equation for the
system but the difficulty with this procedure is soon apparent: the
distance between the electrons, r_{12}, involves the position coordinates
of both electrons *simultaneously*. All the other distances (Figure 3.1)
are only electron–nucleus distances (or the internuclear separation).
Because of the term r_{12} it is impossible to solve the wave equation
in the 'exact way' and so it is no use trying.

This situation need not cause disillusion because the problem is
not new even in elementary mathematics. The equation $x^2 - 2x
- 8 = 0$ is easily soluble by factorization yielding $x = -2$ and
$x = +4$ but the equation $x^2 - 2 \sin x - 8 = 0$ is not soluble by
simple factorization or, indeed, by any 'exact' method. It *does*
however have solutions, e.g. if we substitute $x = 2$ radians we get
$x^2 - 2 \sin x - 8 = -5.84$, whereas if x is put equal to 3 radians
the whole expression $= +0.68$. To refine the answer we would now
try $x = 2.7$ or 2.8 radians. The same principles as operate in the
algebraic example apply equally well to our choice of a wave function for the hydrogen molecule, i.e. we first of all make a guess at it.
It is reasonable that the H_2 molecular wave function should be
related in some way to the hydrogen $1s$-orbital itself. A possible
(guessed) form is the product of two normalized $1s$-orbitals thus

$$\Psi_{H_2} = \psi_A(1) \times \psi_B(2).$$

It must be stressed that the form of this function is pure guesswork –
another, quite different, form will be introduced in the next chapter.
The function is just a simple, convenient, mathematical form and

using it is like assuming $x = 1$, 2 or 3 in the insoluble algebraic equation above rather than substituting some exact number like $x = 1 \cdot 47254$, for instance. The principal difference in the H_2 situation is that we are testing the fit of a *function* to a differential equation rather than numbers in an algebraic one.

A second difference is that subsequent alterations to the above wave function to get a better fit are guessed, not haphazardly, but in a rather more systematic way. The best way to judge the accuracy of a guessed wave function is not to fit it in the Schrödinger equation and examine the result (as we did with the trial solutions of the algebraic equation) but to substitute it into the integrals of the energy equation (3.1) and judge how well it reproduces experimental results. If the above 'product' form of the hydrogen molecular wave function is employed in this way then precisely the same form of integrals results as was obtained by the earlier quasi-physical reasoning. It is worth showing this in detail for one of the integrals before passing on to the next argument.

Consider one of the original integrals

$$\int \frac{\psi_A(1)^2}{r_{A1}} \, d\tau_1$$

which does not contain $\psi_B(2)$ at all. If the product wave function (i.e. for the molecule H_2) is substituted for $\psi_A(1)$ a new form is obtained,

$$\int \int \frac{\psi_A(1)^2 \, \psi_B(2)^2}{r_{A1}} \, d\tau_1 \, d\tau_2.$$

Here there are two integral signs because the integrations are over $d\tau_1$ and $d\tau_2$, i.e. the 'spaces' of electrons (1) and (2). This integral may now be rewritten, as,

$$\underbrace{\left\{ \int \psi_B(2) \, \psi_B(2) \, d\tau_2 \right\}}_{A} \int \frac{\psi_A(1)^2}{r_{A1}} \, d\tau_1.$$

(Remember that r_{A1} is the distance from electron (1) to nucleus A.) Since $\psi_B(2)$ is a normalized $1s$-function, the integral $A = 1$ and the whole contracts to

$$\int \frac{\psi_A(1)^2}{r_{A1}} \, d\tau_1$$

just as was derived physically in expanding eqn (3.1). Using this form of wave function the kinetic energies of the electrons are also precisely the same for the molecule as for two isolated hydrogen atoms, and so do not contribute to molecular stability. If all the integrals of eqn (3.1) are evaluated for a series of internuclear distances, R, using the product wave function, an energy curve similar to that in Figure 3.2 is produced. Its minimum is 6 kcal mole^{-1} lower in the energy scale than the two separated atoms. This shows that two associated hydrogen atoms are indeed more stable than two separated ones, but when the predicted energy of combination is compared with the experimentally observed energy of dissociation ($\sim$110 kcal mole^{-1}) the agreement between the two is obviously very poor.

The principal problem is solved, however, and all that we now have to do is reduce the discrepancy. This can be done by improving the wave function. Now our adoption of the simple product function $\psi_A(1)\ \psi_B(2)$ is equivalent to supposing that electron (1) resides on nucleus A and electron (2) on nucleus B. In the separated atoms this is certainly true but in the molecule the possibility that electron (1) can wander onto nucleus B and vice versa ought to be taken into account. This is expressed by adding the new term $\psi_A(2)\ \psi_B(1)$ thus producing a new function, i.e.

$$\Psi_{H_2} = \frac{1}{\sqrt{2}}[\psi_A(1)\ \psi_B(2) \pm \psi_A(2)\ \psi_B(1)]. \qquad (3.2)$$

Although the individual ψ's are normalized, the factor $1/\sqrt{2}$ in front is necessary to normalize the *whole* combination. This type of wave function, which is simply a sum or difference of terms, is called a *linear combination*. When taking a linear combination it is always necessary to put in the normalizing factor even if the component terms are themselves normalized (see end of chapter for a problem). In the second term the electron numbering is interchanged and this corresponds to 'delocalization' of the electrons. Little importance should be attached to the *physical* idea of interchange of electrons; the term has really arisen because electrons (1) and (2) are indistinguishable. The new wave function may now be tried out in one of the 'test' attraction integrals to see what difference the added term makes. We only need treat one integral because a similar effect will operate for the others. For the linear combination function with the *positive* sign the modified expression is (leaving aside the $1/\sqrt{2}$ factor),

$$\text{Attraction energy} = \int \int \frac{[\psi_A(1)\,\psi_B(2) + \psi_A(2)\,\psi_B(1)]^2}{r_{A1}}\,d\tau_1\,d\tau_2 \quad (3.3)$$

which on expansion gives,

$$\text{A.E.} = \int \int \frac{\psi_A(1)^2\,\psi_B(2)^2 + 2\psi_A(1)\,\psi_B(2)\,\psi_A(2)\,\psi_B(1) + \psi_A(2)^2\,\psi_B(1)^2}{r_{A1}}\,d\tau_1\,d\tau_2.$$

Each separate integral in the brackets may be factorized thus:

$$\text{A.E.} = \int \underbrace{\psi_B(2)^2\,d\tau_2}_{A} \int \frac{\psi_A(1)^2}{r_{A1}}\,d\tau_1 + 2\int \frac{\psi_A(1)\,\psi_B(1)}{r_{A1}}\,d\tau_1$$

$$\times \int \underbrace{\psi_B(2)\,\psi_A(2)\,d\tau_2}_{B} + \int \underbrace{\psi_A(2)^2\,d\tau_2}_{C} \int \frac{\psi_B(1)^2}{r_{A1}}\,d\tau_1 \quad (3.4)$$

Factors A and C occurring in the first and last terms are equal to unity because of normalization and the first and third terms have been encountered previously. The most interesting term is the middle one which is divided into two parts. The first is a sort of 'mixed attraction integral' between electron (1) and nucleus A. The second factor, $\int \psi_B(2)\,\psi_A(2)\,d\tau_2$, is extremely important. It involves one electron (number (2)) and *both* nuclei A and B. Because the nuclei lie at the *two* centres of coordinates we cannot simply make the integral equal to unity, as would be true for a one-centre integral. The integral has a finite value, which depends on the distance between A and B, and is called the *overlap integral*. The name arises from the form of the integral – it represents the overlapping of the two functions $\psi_A(2)$ and $\psi_B(2)$ over the distance R. This is a very important quantity in chemical applications and its meaning will presently be discussed in some detail. When the new form of wave function, i.e. the linear combination with the plus sign, is substituted into the terms of eqn (3.1) and the value of each worked out, the energy of the system drops to a deep minimum at the equilibrium point. The dissociation energy is now ~72 kcal mole^{-1} which is much nearer to the experimental value. There is, of course, a second linear combination (with a minus sign) which has to be tested. The integrals can be re-evaluated using this wave function and, when this is done for a set of distances R, the energy curve *rises* from the initial energy of two isolated atoms and has no minimum at all. This means that this wave function does not describe a bonding situation between the two atoms. The important point which has been established is that modification of the wave function by turning it into a linear combination is able to bring us closer to finding the true energy of the hydrogen molecule.

Heteropolar Diatomic Molecules, and the Overlap Integral

The present wave function for the system embodies two terms: the first represents electron (1) being attached to nucleus A and electron (2) to nucleus B. The second term represents the opposite arrangement. Both, however, visualize each nucleus as possessing an electron and because of this they are called *covalent* terms. Further possibilities could arise if both electrons were located on nucleus A or both were on nucleus B. To take account of these we add in the new terms,

$$\psi_A(1) \times \psi_A(2) \quad \text{and} \quad \psi_B(1) \times \psi_B(2).$$

Because these terms represent a separation of charge (the physical situation H^+H^- or H^-H^+) they are called *ionic* terms. Their inclusion in the wave function increases the stability of the system and brings the calculated energy yet nearer to the experimental value. It should be obvious that in a homopolar molecule like hydrogen these ionic terms will not be particularly important; chemical intuition tells us that hydrogen is a 'mainly-covalent' molecule. It is interesting that all further reasonable modifications made to the wave function (and there are a number of these) always add to the calculated stability of the molecule though usually to a systematically decreasing extent. There is, in fact, a theorem governing the procedure (as might have been guessed); it says that *the true energy of the system is always the lowest of all* so that if we start with a crude approximation to the wave function and improve it step by step we shall gradually approach the true answer. The theorem is called the *Variation Principle*. There are a number of mathematical techniques attached to the usage of this theorem but we shall not trouble with them; it is the idea itself which is most important.

The ionic terms in the wave function merit further consideration. They may not appear to be very important in a homopolar molecule like hydrogen but are obviously going to assume far greater prominence when the bond under consideration is heteropolar, e.g. that in H—F. On general chemical grounds it is known that fluorine is much more electronegative than hydrogen and so terms representing H^+F^- contribute far more strongly to the overall wave function. To generalize, it is convenient to write the *total* wave function of a diatomic system as,

$$\Psi_{total} = \psi_{covalent} + a\psi_{ionic}. \tag{3.5}$$

Here $\psi_{covalent}$ incorporates all the covalent terms and ψ_{ionic} all the appropriate ionic terms which compose Ψ_{total}. If the molecule is homopolar or only weakly heteropolar then 'a' will be small but if strongly heteropolar 'a' can be quite large. The practice of guessing at the magnitude of 'a' (which we can do even at this early stage) has been one of the most valuable facets of valence-bond theory and at the same time one of its chief pitfalls. The chemist, using his intuition, can suggest which type of term is likely to make the most telling contribution to the overall wave function. In fact this is just what we have been doing in an elementary way with the molecules H_2 and H—F. The magnitude of 'a' represents in some measure the *degree of ionic character* of a bond and could be conjectured from chemical experience, e.g. how readily the compound dissociates electrolytically in water. Or we might choose to employ the *dipole moment* to aid in assessing 'a'; after all, this is an experimental measure of the electrical asymmetry in the molecule. Such correlations look quite good on the face of it and do indeed enable a guess at 'a' to be made.

Unfortunately there are a number of serious deficiencies inherent in using *overall* molecular properties to construct wave functions, e.g. the use of the dipole moment for this purpose looks straightforward and in H—Cl the dipole moment indicates a drift of electronic charge from H to Cl (i.e. H^+Cl^-). We might therefore attempt to get 'a' from its magnitude. However, two important features are neglected by such an approximation. To make these clear first suppose that the two atoms in hydrogen are drawn as circles (Figure 3.3 (*a*)). The overlap charge is then midway between the two atoms and it is easy to understand why this molecule has no dipole moment. If HCl is drawn similarly (Figure 3.3 (*b*)) and, assuming first that there is *no* charge distortion in the bond, it is clear that the average radius of chlorine is greater than that of

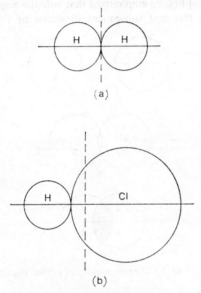

(a)

(b)

FIGURE 3.3. Origin of the homopolar dipole

hydrogen. Hence the overlap charge (if there is no charge distortion) will not be centrally located between the two atoms (at half the bond length) but will be displaced towards hydrogen. A dipole moment will therefore be manifest even *before* taking account of charge drift. Such a dipole is called a *homopolar* dipole.

Secondly the chlorine atom in HCl possesses a number of other *lone-pairs* of electrons as well as the one which takes part in bonding. These make a contribution to the dipole moment which is difficult to assess and so subtract out. What is wrong with employing the dipole moment to guess 'a' is not that it is intrinsically a bad idea but that it is a physical property which is determined by other factors beside the bond charge asymmetry.

It is now convenient to return and discuss the properties and meaning of the overlap integral. Firstly, it has strong dependence on the internuclear distance, R. Figure 3.4 illustrates three situations involving different degrees of overlap between two $1s$ atomic orbitals. It must first be emphasized that *only* the angular functions are drawn in this and subsequent diagrams of the same kind.

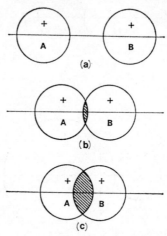

FIGURE 3.4. Representation of orbital overlap

Strictly, this is not sufficient because physical distances between atoms have no meaning with regard to the angular functions alone. However, as all radial functions incorporate exponential terms they may only be drawn as contour diagrams. This complicates considerably a diagrammatic study of overlap and so at all stages it must be remembered that the radial part of the total wave function has been neglected and that the diagrams considered only contain this function implicitly.

In Figure 3.4 (*a*) the atoms are far apart and the overlap of the two functions is very small: Figure 3.4 (*b*) shows greater overlap and the overlap integral is small, but finite (represented diagrammatically by the shaded portion on the figure). In Figure 3.4 (*c*) the overlap is still further increased. If the two nuclei are coalesced so overlap is a maximum, the two functions ψ_A and ψ_B are then based on the *same* axis system and the overlap integral is unity just as if one function was simply normalized.

What relation does the overlap integral bear to the bonding of two atoms? We know that it depends on the distance apart of the two atoms, e.g. at infinite distance apart it is zero and consistent with this we know from experience that there is no bonding when the atoms are separated by an infinite distance. Thus we would expect the overlap integral to be related directly in some way to the *degree of bonding* between two elements. (This conclusion is reasonable because in eqn 3.4 the integral appears as a factor which multiplies a stabilizing energy term – the so-called 'exchange term' – so that this term at least must be dominated by it.) There is a reservation to this though: the strengths of bonds may only be compared *via* the overlap integral of the orbitals involved when we are talking about the same bond, e.g. we could say, using overlap data, that an H—H bond of x Å length was twice as strong as one of y Å in length, but we could not say that the bond in H—Cl was twice as strong as that in H—Br because the calculated overlap integral between the relevant orbitals was twice as large for the former.

Let us now consider the valence-bond picture of a molecule such as HF. It is represented by the overlap of a hydrogen $1s$-orbital containing one electron with a $2p$-orbital of fluorine, also with one electron (the atomic structure of fluorine, $Z = 9$, is $1s^2 2s^2 2p^5$). There are many ways in which the fluorine $2p$-orbital could be orientated with respect to the $1s$-orbital – three of these are shown in Figure 3.5. In (*a*) the *p*-orbital points directly towards the hydrogen atom along the line of atomic centres; in (*c*) the *p*-orbital is directed at *right angles* to the line of centres. These represent two limiting cases and Figure 3.5 (*b*) is one of the infinite number of possibilities lying in between. Which of these will be favoured? Case (*c*) may be ruled out immediately because the net overlap is zero. This results from cancellation of the 'positive overlap', above the dividing line by the equal 'negative overlap' below it. (There are rather more elegant reasons than this for the vanishing of overlap and these will be investigated in a later chapter.) Two orbitals

72   ELEMENTARY MOLECULAR BONDING THEORY

orientated with respect to each other in this way so that the net overlap is zero are said to be *orthogonal* and they do not bring about bonding between the respective atoms. Following from

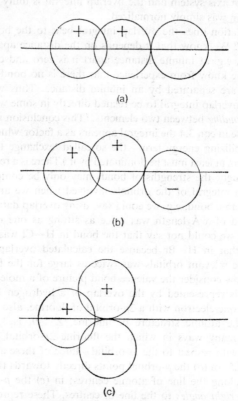

FIGURE 3.5. Possible overlap between H $1s$ and F $2p$

this it should now be obvious that for a given internuclear distance orbital overlap will be greatest in case (*a*); this is because the functions drawn are angular functions and the maximum extension of a *p*-orbital is along its polar axis. In the two bonds we have dealt with, i.e. $1s$–$1s$ in H_2, and $1s$–$2p$ in HF, overlap, and therefore bonding, occurs along the line of atomic centres in each case. This type of bond is called a *sigma bond* by analogy with the *s*-orbital.

Figure 3.6 (*a*) illustrates another type of σ-bond and could represent the $2p$–$2p$ F—F bond in F_2. Here both *p*-orbitals are directed

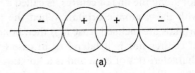

(a)

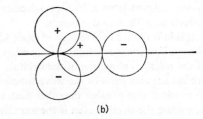

(b)

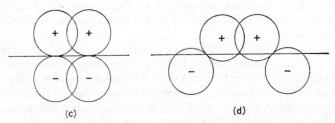

(c)   (d)

FIGURE 3.6. Possibilities for *p*-orbital overlap

towards each other along the line of centres. The same description as in F—F would also apply to the Cl—Cl or Br—Br bonds although for the latter the $3p$–$3p$- and $4p$–$4p$-orbitals would be employed. (Remember that *p*-angular functions are identical independent of their principal quantum number.) Case (*b*) is interesting because the overall orbital overlap is zero and this is again due to the cancellation of 'positive' and 'negative' overlap above and below the dividing line. Case (c) is different again; here the overlap is not now along the line of centres (in fact it is zero in this direction) but nevertheless there is still *genuine* overlap which leads to bonding in a lateral fashion. The kind of bonding thus brought about, which is not along the line of centres, is called a π-bond, this time by analogy with the p atomic orbital. When the polar axis of one (or both) orbitals are inclined (case (d)) then bonding is neither along the line of centres nor is it lateral as in case (c). In fact it possesses some characteristics of each and is a 'mixture' of both a σ- and a π-bond. Thus, like a vector, it can be resolved into these two components. We ought to add here that case (d) will never arise in a consideration of bonding in diatomic molecules but only in some polyatomic systems. Apart from a few special cases the chemist need only be concerned with σ- and π-bonds.

It is informative here to inspect cases 3.5 (c) and 3.6 (b) a little more closely. There was resultant zero overlap in these situations because of the *relative orientation* of the orbitals. Now the orbitals depicted are the angular parts of the wave function *alone* and it is necessary to consider also whether the other part, i.e. the radial function, can annihilate the overlap even if the angular functions do not. Clearly it can, because if the atoms are separated by infinite distance then the angular functions still overlap as they contain no distance variable but the radial functions do not because of the ∞ atomic separation.

It will now be worthwhile to summarize the results of our study of the overlap problem. In the bonding of two atoms the constituent orbitals must overlap to some extent. This means at (i) the orbitals must be directed towards, or be parallel to each other, (ii) the atoms must lie reasonably close together. The former condition means that, however close the atoms approach, there will be no overlap (or bonding) *unless* the orbitals are orientated correctly: the second means that even if the angular orientation is favourable there will be no bonding if the atoms are separated physically by large distances.

We must bear in mind constantly that the functions depicted in Figures 3.5 and 3.6 are *angular* functions and only tell us about the probability or electron density in a given *direction*. Hence the distance apart of the atoms does not affect in the slightest the relative orientation of the angular functions and although their radial functions may not overlap since these *do* depend on distance, we can always draw diagrams like Figures 3.4 and 3.5 to represent bonds. The converse is perhaps worth comment: just because these bonding pictures may be drawn does not mean that orbital overlap occurs to an extent capable of explaining all electronic effects in molecules – only by estimating the *total* overlap can this be done.

Now it was stated that, in the context of one particular bond, increased overlap leads to increased bonding and stability and, moreover, overlap increases to unity when the atoms coalesce. Why then does not the latter always happen to produce the strongest possible bonding? This question is readily answered by referring back to eqn (3.1). The overall equilibrium situation in a bond is not due just to overlap but is the result of the interplay of a number of factors which are both attractive and repulsive; overlap influences only certain of these. One might as well ask why atoms bond together at all because both electrons and nuclei repel each other.

On similar lines it is often supposed that the shapes and bond lengths in molecules are determined by electron repulsion forces alone. This argument again neglects a large proportion of the interactions because if electron repulsions alone mattered, no molecules would be formed at all. The characteristics of all molecules result from a compromise between the attractive (stabilizing) forces and the repulsive (destabilizing) forces. Of course, frequently one of the factors may be picked out to dominate a particular situation.

Application of Valence-Bond Theory to Polyatomic Molecules

Our study of hydrogen started from the supposition that the hydrogen atoms were, initially, an infinite distance apart and had each an unpaired electron. Valence-bond theory assumes that, if two orbitals are to overlap and bring about bonding between atoms, then they require to have either an unpaired electron each or one of them has to contain two electrons (as in the coordinate bond). These criteria bear a distinct relation to the Lewis electron-pair bond theory and this is why valence-bond theory was (and still is) so readily assimilated and used by chemists. It carries into more sophisticated terms the earlier idea of bonds being 'sticks' between the atoms. The requirement of two initially unpaired electrons for simple covalent bond formation is one of the salient points of the theory and much of what now follows was proposed to achieve this end. In order to develop the point the valence-bond picture of diatomic molecules should be extended to cover larger systems containing mixed orbital types. One of the most fascinating features of polyatomic molecules is that they exhibit a variety of shapes and so clearly it will be necessary to consider overlap in different directions also. One of the primary questions posed in Chapter 1 was why molecules do take up such shapes and the problem can be attacked immediately by application of valence-bond theory. It is best to consider a polyatomic molecule as being broken down into its diatomic components, i.e. we take the system bond by bond and consider the overlaps appertaining to each case. Two familiar examples are H_2O and NH_3.

Water

The oxygen atom has the electronic structure $1s^2 2s^2 2p^4$ (Figure 3.7 (*a*)) in which the triply degenerate *p*-orbitals contain four elec-

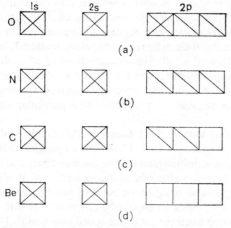

FIGURE 3.7. Electronic configurations for oxygen, nitrogen, beryllium and carbon

trons and so two orbitals containing each an unpaired electron are available. These two unpaired electrons are contained in two *p*-orbitals which are orientated at right angles to each other on the oxygen atom (Chapter 2) and so, if each p-orbital overlaps with the $1s$ of a hydrogen atom, the water molecule may be drawn as in Figure 3.8. This description of the bonding in water predicts that the molecule ought to be angular with an H—Ô—H angle of 90°. Before commenting further on this result it is instructive to examine the next molecule, NH_3. The hydrogen atoms use the same $1s$-orbitals as in water but nitrogen differs from oxygen in its electronic structure (Figure 3.7 (b)) – here all three p-orbitals are occupied by unpaired electrons. This allows three N—H bonds to be formed by overlap of the three $2p$-orbitals with the three hydrogen $1s$. Moreover the

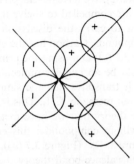

FIGURE 3.8. Orbital overlap in H_2O

three bonds should be mutually inclined at right-angles and a pyramidal shape would be predicted for ammonia. In both NH_3 and H_2O these bonds are of σ type. This is necessarily true because s-orbitals (on H in these cases) cannot participate in π-bonds.

In this treatment each bond has been considered individually and overlapping of the relevant orbitals carried out separately just as if the system were an assemblage of diatomic units. The reason for (or mechanics of) the formation of the two OH and the three NH bonds does not differ in the slightest from the example of H_2 already examined. Of course, these systems are more complex than H_2; the orbitals on each atom are not of the same type nor are the nuclear charges equal. The two H nuclei in H_2O will also repel each other, just as their $1s$-orbitals will, to a certain extent, overlap and bond even though the distance between the atoms is rather large. The latter factor will emerge more prominently in Chapter 5 and will be discussed there.

A difficulty with our formulation of H_2O and NH_3 is encountered when the results are compared with experiment. The predictions turn out to be *generally* true: H_2O *is* an angular molecule, and NH_3 *is* pyramidal not planar. However, in neither case are the angles between the X—H bonds equal to 90° as we would have expected. Does this mean that the orbital overlap does not take place directly along the line of atomic centres? This is surely not true because the *maximum* directional probability of finding an electron is along the coordinate axes (Chapter 2). We have developed no theory to help us solve this problem so are compelled to shelve it for the time being. Instead let us attempt to carry the discussion further and study carbon–hydrogen compounds. Carbon has the electronic configuration $1s^2 2s^2 2p^2$ (Figure 3.7 (*c*)) and using the principles outlined, the simplest hydride should be an angular molecule, CH_2, rather like H_2O. Experimentally it transpires that the simplest stable hydride of carbon is not CH_2 but methane, CH_4. This is the first qualitative disagreement with the theory that we have met. Further examples are furnished by beryllium compounds: this element has ground state electronic structure $1s^2 2s^2$ (Figure 3.7 (*d*)), i.e. it possesses *no* unpaired electrons. On valence-bond theory therefore it would be expected to have a 'valency' of zero and hence not to give rise to any chemical compounds. Of course, many compounds of beryllium are known and obviously the theory must be modified to accommodate these, and other, 'awkward' cases.

Hybridization

The difficulty may be circumvented in the Be case if an electron is 'promoted' to one of the higher-energy, empty, $2p$-orbitals giving rise to the configuration Be $1s^2 2s^1 2p^1$ (often designated Be*). Such a configuration incorporates *two* unpaired electrons and bonding to two other atoms having each an unpaired electron (e.g. two fluorine atoms) can be explained. However, the linear symmetry of monomeric BeF_2 (i.e. two equivalent bonds) is left unexplained because it might be expected that the overlap between one fluorine $2p$-orbital and the $2s$-orbital of Be would be directionally different to the other involving the Be and F $2p$-orbitals. To overcome the discrepancy we suppose that the s- and one p-orbital of Be* are 'mixed' to form two new ones which are equivalent and directed in a somewhat different way to the old. This orbital mixing is called *hybridization* and it may readily be illustrated pictorially, provided that the radial part of the orbital is neglected. This is an approximation but it is worth making for the understanding of the situation which results.

Hybridization is in no sense a physical event: orbital 'mixing' is only a mathematical process. What we do is to construct two new hybrid orbitals by taking the two possible algebraic linear combinations of the old, i.e.

$$\frac{1}{\sqrt{2}}(s + p_z)$$

and
$$\frac{1}{\sqrt{2}}(s - p_z) \quad \text{(defining the Be–F axis as the z-axis)}$$

The factor $1/\sqrt{2}$ is the normalization factor, necessary for the same reasons as were discussed previously. The shapes of the new hybrid orbitals may readily be derived as follows using only the angular functions. Let the radius of the s-angular function be unity; relative to this the p_z-function has maximum extension along the z-axis equal to $\sqrt{3}$ (Chapter 2). These are *not* distances in physical space but only on the function itself, i.e. the graph. Next for a set of angles α (measured from the z-axis) the values of the two functions along that direction for the hybrid $s + p_z$ are added together. Table 3.1 lists this sum (multiplied by $1/\sqrt{2}$) for a set of angles α from $0°$ to $360°$ and Figure 3.9 (*a*) illustrates the plot of $(1/\sqrt{2})$ $(s + p_z)$ for $360° > \alpha > 0°$. The author strongly recommends to the reader the exercise of calculating and plotting the appropriate graphs. The technique is very similar to that described in Chapter 2 for atomic angular functions. Normalization of the hybrid is merely a technicality and does not affect the force of the chemical arguments. The second combination, $1/\sqrt{2}$ $(s - p_z)$, may be calculated and plotted by the same method; we shall not carry this out, but it is easy to see that a similar diagram will be obtained, but orientated at $180°$ to the first (Figure 3.9 (*b*)). (The maximum extension of each along the z-axis, relative to an s-orbital, is $1·932$.) The procedure therefore produces two new equivalent functions which are

Table 3.1 *sp–Hybridization*

α (deg)	s	p_z	$\frac{1}{\sqrt{2}}(s + p_z)$
0	+1	+$\sqrt{3}$	+1·932*
30	+1	+1·499	+1·768
60	+1	+0·866	+1·319
90	+1	0	+0·707
120	+1	−0·866	+0·095
150	+1	−1·499	−0·353
180	+1	−$\sqrt{3}$	−0·518
210	+1	−1·499	−0·353
240	+1	−0·866	+0·095
270	+1	0	+0·707
300	+1	+0·866	+1·319
330	+1	+1·499	+1·768
360	+1	+$\sqrt{3}$	+1·932

* This is the maximum extension of normalized *sp* hybrid relative to an *s*-angular function.

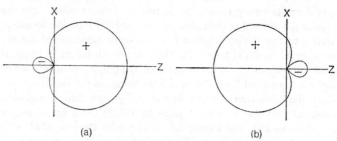

FIGURE 3.9. Digonal (*sp*) hybrid functions

differently shaped from the original *s* and *p* types and which lie at 180° to each other. These are called *digonal* or *sp* hybrids.

There are several points worth noting at this juncture because they are the root cause of many common misconceptions in the application of the ideas of hybrid orbitals.

(i) Only a *mathematical* mixing of the orbitals is brought about and this *cannot be equated to a physical process*.

(ii) The extension of the hybrid angular function along the *z*-axis is greater than that of either the *s*- or *p*-functions.

(iii) The surface of the function intersects the x- and y-axes and has the value $+1$ in these directions.

(iv) The promotion of an electron to the higher level is only a hypothetical process and is not something which 'happens'.

(v) *Two* initial atomic orbitals gave rise to *two* hybrids. (This is generally true and could be called the 'law of conservation of orbitals'.)

These points will be considered in detail shortly.

A second case meriting consideration is that of boron with ground-state configuration $1s^2 2s^2 2p^1$. As before, the valency problem may be overcome by postulating attainment of the state, B^* $1s^2 2s^1 2p_x^1 2p_y^1$ by promotion of a $2s$ electron. The observed trigonal symmetry of boron compounds may then be explained on the basis of hybrids formed by 'mixing' the $2s$, $2p_x$, and $2p_y$ orbitals which, by (v) above,

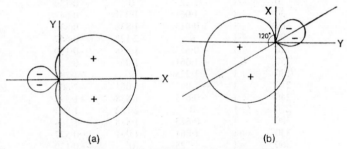

FIGURE 3.10. Trigonal (sp^2) hybrid functions

must yield *three* new hybrid orbitals. We may derive their directional properties as before from the form of the linear combinations of the angular functions. These are

$$h_1 = \frac{1}{\sqrt{3}}(s + \sqrt{2}p_x)$$

$$h_2, h_3 = \frac{1}{\sqrt{3}}\left(s - \frac{p_x}{\sqrt{2}} \pm \frac{\sqrt{3}p_y}{\sqrt{2}}\right).$$

From the first combination a hybrid similar to the sp may be derived and it may readily be shown that its normalized maximum extension is $1\cdot991$ (cf. $1\cdot932$ for sp), (Figure 3.10 (*a*)). The other two hybrids are interesting: they involve *both* p_x and p_y simultaneously and because of this they will not be directed along any particular Cartesian axis, but somewhere in the x, y plane. It is instructive to plot one of these (e.g. h_2, with the $+$ sign for p_y) and the necessary figures are listed in Table 3.2.

On plotting out the linear combination proportional to the angle α we obtain Figure 3.10 (*b*). The angle between the polar axis of the hybrid and the positive Cartesian x-axis is 120° and, moreover, as in Figure 3.10 (*a*), the hybrid intersects the y-axis. It is fairly obvious

Table 3.2 sp^2 Hybridization

α (deg)	s	$p_x/\sqrt 2$	$\sqrt 3 p_y/\sqrt 2$	Hybrid
0	$+1$	$+1\cdot225$	0	$-0\cdot130$
30	$+1$	$+1\cdot061$	$-1\cdot061$	$-0\cdot647$
60	$+1$	$+0\cdot613$	$-1\cdot838$	$-0\cdot837$
90	$+1$	0	$-2\cdot122$	$-0\cdot647$
120	$+1$	$-0\cdot613$	$-1\cdot838$	$-0\cdot130$
150	$+1$	$-1\cdot061$	$-1\cdot061$	$+0\cdot577$
180	$+1$	$-1\cdot225$	0	$+1\cdot284$
210	$+1$	$-1\cdot061$	$+1\cdot061$	$+1\cdot802$
240	$+1$	$-0\cdot613$	$+1\cdot838$	$+1\cdot991*$
270	$+1$	0	$+2\cdot122$	$+1\cdot802$
300	$+1$	$+0\cdot613$	$+1\cdot838$	$+1\cdot284$
330	$+1$	$+1\cdot061$	$+1\cdot061$	$+0\cdot577$
360	$+1$	$+1\cdot225$	0	$-0\cdot130$

* This is the maximum extension of the normalized hybrid relative to an s-angular function.

that the third hybrid, h_3, must lie in the plane at 120° to the first two and thus three equivalent *trigonal* or sp^2 hybrids are produced.

All the principles required to discuss the case of carbon are now available. In this case the ground-state configuration is $1s^2 2s^2 2p^2$ and C* has the configuration $1s^2 2s^1 2p_x{}^1 2p_y{}^1 2p_z{}^1$. Mixing of *all four* of these orbitals leads to *four* equivalent sp^3 hybrids, the angular parts of which are directed to the corners of a regular tetrahedron. The graphing of the hybrids could, in principle, be carried out as befroe, but for a tetrahedron the geometry is rather more complex.

The difficulty here lies in trying to represent a three-dimensional diagram on two-dimensional paper. Each total hybrid involves a symmetrical linear combination of the s- and *all three* p-orbitals and has normalized maximum polar extension 2·00. The forms of the new hybrid orbitals are

$$h_1 = \tfrac{1}{2}(s + p_x + p_y + p_z)$$
$$h_2 = \tfrac{1}{2}(s + p_z - p_y - p_z)$$
$$h_3 = \tfrac{1}{2}(s - p_x + p_y - p_z)$$
$$h_4 = \tfrac{1}{2}(s - p_x - p_y + p_z).$$

What is possibly most interesting and instructive about the hybridization approach is that the observed symmetries, in limiting situations, of molecules under discussion (e.g. BeF_2, *linear*, BF_3, *trigonal*, and CH_4, *tetrahedral*) 'arise' quite naturally which lends an air of 'prediction' to what one is doing.

The points mentioned earlier, which can lead to misunderstanding, will now be considered in more detail.

(i) The mechanics of hybridization

The mixing of orbitals in hybridization is a mathematical process and it cannot be linked in any way to a physical one. This should not be surprising because, after all, an s-orbital and a p-orbital are mathematical functions and not real physical objects. The chief way in which hybridization loses value in the understanding of molecular bonding is when one attempts to tie it too closely to physical phenomena. What has been done is to facilitate *description* of the bonding in a molecule by mixing the orbitals on one atom which are to be overlapped. It is therefore ridiculous to imagine that, before forming methane, the carbon atom must 'become hybridized', i.e. a kind of physical process must take place as some essential step in a reaction sequence. This leads to the rather incongruous picture of a carbon atom sitting with its four hybrid orbitals outstretched waiting for four hydrogen atoms to approach and attach themselves. It also means that we must be prepared to say how long such a 'process' will take to complete – do atoms which approach before this stage have to retreat and return later? How does hybridization actually take place? Does the atom 'wriggle' to mix the orbitals? These questions seem ludicrous and in fact none of them is relevant unless we believe that hybridization is a physical event. Thus, although it is common, and perhaps harmless jargon, it is also inappropriate to say 'in compound X the metal atom is sp^2 hybridized'. Hybridization is not a physical process nor indeed an *essential* 'process' of any kind, as will become more evident when the approach to bonding through molecular orbital theory is made. This does not require the postulation of hybrid orbitals at all.

(ii) Hybrid extensions

The maximum extension on the graph of the hybrids along the poles is different for the sp, sp^2 and sp^3 cases and is always greater than either an s- or p-orbital alone. In this fact lies a wide-open trap. If we accept that bonding and overlap are interrelated, it is reasonable to expect that greater overlap will lead to stronger bonds. Hence it is tempting to consider that the greater extension of the angular part of a hybrid allows larger overlap along the bond direction and consequently a stronger bond. This reasoning has two failings, one of which is mathematical and one logical. They are (a) the hybrids were built from angular functions alone and so take no account of the radial function and its associated distance variable. We cannot simply neglect this because the orbital overlap may well be zero at specified interatomic distances (infinity is one obvious one). Therefore care must be exercised in citing the angular function alone as a criterion of bond strength. This argument of course applies whether or not we consider the orbitals hybridized.

Perhaps a more significant point is that (b) molecular orbital theory does not introduce the concept of hybridization at all, yet *precisely the same overall overlap* may be calculated for a given limiting case on the basis of separate unhybridized orbitals. This is sensible because otherwise we could account for a molecule apparently satisfactorily by m.o. theory and then cause the bonds to become stronger by merely switching to an alternative hybrid description of it! An impression is often gained that the atom must 'get something out of' hybridization or it wouldn't 'do' it and then it is suggested that greater overlap is achieved by hybridization leading to stronger bonds. Such a view is untenable.

(iii) Excited states

There is frequently much confusion about what is meant by the 'excited' state of the atom (i.e., the valence state M*). It may seem that energy has to be supplied from some source in order to 'excite' the atom to this state before hybridization can 'take place', and presumably, before the atom can react with a second entity. It also seems plausible to construct a specific excited state e.g. $C^* 1s^2 2s^2 2p_x{}^1 2p_y{}^1 2p_z{}^1$, in which there are four unpaired electrons. This is often supposed to be a definite state of the atom (designated as ^5S) in which all four electrons have the *same spin* and which lie ~ 95 kcal above the ground state. Such an assumption is unjustifiable because if an electron really was physically promoted to attain the valence state we would not know whether it, or the ones remaining, had α or β spin. Hence we cannot say that the ^5S state (a quintet) is the valence state; there are a number of other possible states of the carbon atom which possesses four unpaired electrons. In fact, the valence state is a *hypothetical* state corresponding to *random electron spins* and is therefore a *mixture* of different spin states. This alone ought to be a convincing argument against the real existence of the valence state. All these difficulties melt away if it is realized that what valence-bond theory does is to *describe* the chemical situations presented to it. They were, after all, there before the theory. Orbital theories describe the behaviour of electrons in atoms and molecules and hybridization is an addend to this description introduced to explain the observed symmetries of molecules. Electrons are not promoted in the sense of being physically moved from orbital to orbital but only *formally* unpaired to produce the appropriate number of unpaired electrons required by valence-bond theory. If we try to adhere to the idea that 'whole' electrons have to be 'promoted' then it will be found that another range of situations cannot be understood which do not accede to the simple rules generated by the limiting situations described till now.

Hybrids involving Orbitals Other than s and p

Sulphur, like oxygen, is generally considered to have a 'valency' of two. However, it also forms a number of compounds (e.g. SF_4, SF_6) in which it is bonded to more than two other groups. Sulphur hexafluoride is a very stable compound and its shape is that of a regular octahedron (Figure 3.11 (a)), i.e. all the S—F bonds are equal in length.

The structure may be explained satisfactorily on the basis of hybridization in which d-orbitals are utilized. The ground-state configuration of sulphur is $1s^22s^22p^63s^23p^4$, which like oxygen, has only two unpaired electrons. The next empty orbitals energetically available to receive 'promoted' electrons are the 3d set and we may write the excited configuration, S* $1s^22s^22p^63s^13p^33d^2$, in which the

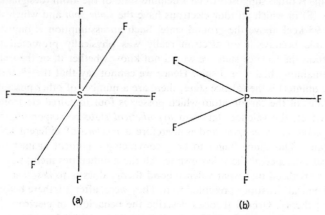

FIGURE 3.11. The shapes of SF_6 and PF_5

'last' six electrons are unpaired and occupy the 3s-, and three 3p- and two of the 3d-orbitals. From these six atomic orbitals, six equivalent hybrids (different linear combinations) may be constructed which involve mixing of s-, p- and d-functions. When these are plotted in three dimensions they extend to the corners of a regular octahedron which corresponds to the observed symmetry of SF_6.

A similar line of reasoning may be adopted to account for phosphorus compounds such as the pentafluorides. The ground-state configuration of the atom is P* $1s^22s^22p^63s^23p^3$ from which 'uncoupling and promotion' of a 3s electron to a 3d-orbital gives rise to P* $1s^22s^22p^63s^13p^33d^1$. Finally, mixing of the five orbitals (s, p and d) containing unpaired electrons leads to five hybrids which are directed to the corners of a trigonal bipyramid (Figure 3.11 (b)). The molecule PF_5 has been shown to possess this structure in the gas phase. A difference between this case and SF_6 is that, whereas all the S—F bonds are equivalent, the P—F are *not* all equivalent and it has been established experimentally that the two axial P—F bonds are longer than the three equatorial ones.

Partial Hybridization

We must now return to rediscuss the observed situation in the water molecule where the measured bond angle is 104° and not 90° as would be expected from simple $1s$–$2p$ overlap criteria. Now we have shown that if two p-orbitals centred on the same atom are mixed with an s-orbital in a predetermined ratio, then three new orbitals (sp^2) mutually directed at 120° are produced. This provides a clue which leads easily to a solution of the 'water-angle' problem: the oxygen p-orbitals which overlap with the hydrogen $1s$ are not pure p-orbitals at all as assumed in Figure 3.8 but include some oxygen $2s$ character, i.e. a proportion of the oxygen $2s$-orbital is mixed in to create 'partial' hybrids. The amount of mixing is related to the actual HÔH bond angle. The OH σ-bond is, as usual, directed along the line of atomic centres. Partial hybridization implies that the $2s$ electron of the oxygen atom is not 'fully promoted' and if we try to fit this in with a physical picture of hybridization (where only integral numbers of promoted electrons can be meaningful) then the mind boggles at what is meant by 'partial promotion'. The situation is easily rationalized if we regard partial hybrids as p-orbitals which have a 'flavour' of s character and *describe* the bonding in the molecules.

The bond angles, and hence the degree of s character in the hybrid, will be governed by all the factors we discussed at length earlier for hydrogen, i.e. internuclear and interelectronic repulsions, overlap criteria. All these will lead to the final compromise situation.

There is a second kind of 'partial hybridization' in which the concept of single electrons promoted also becomes lost. The digonal and trigonal hybrids of Figures 3.9 and 3.10 were generated by plotting *limiting* hybrid functions like the digonal case,

$$h = \frac{1}{\sqrt{2}}(s + p_z).$$

Carrying this a stage further it is important to realize that *any* hybrid form

$$h = \frac{1}{\sqrt{1 + \lambda^2}}(s + \lambda p_z)$$

when plotted out (for a specific value of λ) will yield a hybrid directed along the z-axis, just as in the *sp* hybrid plotted previously. Table 3.3

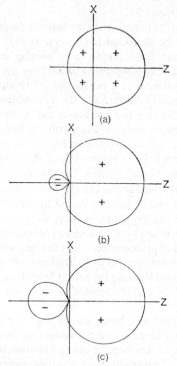

FIGURE 3.12. 'Partial' *sp* hybrids

and Figure 3.12 show the forms of three *sp*-type hybrids generated by taking $\lambda = 0.2$, 1.0 and 2.0. These show an interesting gradation in shape. Figure 3.12 (*a*) is like an *s*-orbital with a bulge on one side – as λ is 0.2 the hybrid contains only a small 'flavour' of *p* character. Case (*b*) is the limiting *sp* case previously discussed and is the only one of the three which correlates with the physical idea of single electron promotion. Case (c) looks much more like a pure p_z-orbital because λ is now large.

The conclusion is quite clear; *there are a whole series* of sp-type hybrid functions lying between the extremes of the 'pure' s- and p-orbitals – and they are *all directed along the z-axis*. There is only one of these which correlates with the promotion of an electron

Table 3.3 *Values of sp: partial hybrid functions*

α (deg)	$\lambda = 0\cdot2$	$\lambda = 1$	$\lambda = 2$
0	1·320	1·932	1·996
20	1·300	1·858	1·902
40	1·241	1·645	1·634
60	1·150	1·319	1·222
80	1·040	0·920	0·716
90	0·981	0·707	0·447
100	0·922	0·494	0·178
120	0·811	0·095	−0·327
130	0·762	−0·080	−0·549
140	0·720	−0·231	−0·739
160	0·661	−0·444	−1·008
180	0·641	−0·518	−1·102

to a p-orbital; the rest are all 'partial' hybrids. Now in any specific case of a compound whose symmetry is explained by digonal hybrids how are we to know what form the hybrid has, i.e. what is the value of λ? The answer is that it will be determined by the nature of the interacting atoms and to assume that λ always $= 1$ (i.e. the limiting case always holds) seems quite unjustified. A similar argument holds for both sp^2 and sp^3 hybrids and similar conclusions may be drawn. However, in the absence of quantitative data the limiting hybrids can always be conveniently assumed.

Organic Molecules

The bonds between atoms in organic molecules can usually be represented by overlap between an atomic orbital and a hybrid or between two hybrids. Hence we may have the following situations

G

(Figure 3.13) which represent the C—H, C—Cl and C—C bonds in CH_4, CCl_4, C_2H_6 respectively.

It should be noticed that, in all cases, the overlap occurs along the line of atomic centres. In Figure 3.13 (a) an sp^3 hybrid orbital of carbon is involved in overlap with a hydrogen $1s$-orbital, in (b) with a chlorine $3p$-function. In (c) the C—C σ-bond in ethane is described by the overlap of two approximately sp^3 hybrid carbon orbitals. Using these principles we can build up many of the familiar molecules encountered in elementary chemistry. All that are needed are the simple hybrids derived from s- and p-orbitals alone.

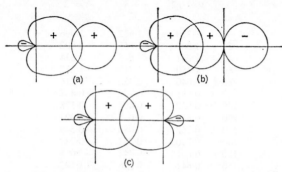

FIGURE 3.13. Hybrid orbital overlaps representing (a) C—H, (b) C—Cl and (c) C—C σ-bonds

Unsaturated organic molecules such as ethylene, C_2H_4, and acetylene, C_2H_2, are quite simply described on the basis of valence-bond theory. A special feature of these two molecules is that all their atoms lie in one plane and to describe them by valence-bond theory the σ skeleton of the system is first built up. Ethylene is known to possess the structure in Figure 3.14 (a) with H—C—H angle near 110°. Mixing of the carbon $2s$- and the two $2p$-orbitals in the plane gives rise to three hybrids. These are actually 'partial' hybrids (like in H_2O) but are very close in form to the trigonal sp^2 type. Two of these overlap with the two hydrogen $1s$-orbitals whilst the third overlaps with a similar 'sp^2' hybrid on the other carbon atom. The two remaining unhybridized p-orbitals on the carbon atoms (each containing one electron) are at right-angles to the molecular plane but parallel to each other. They can thus overlap laterally to form a π-bond (Figure 3.14 (b)). The formulation of ethylene as $CH_2{=}CH_2$ follows reasonably from this valence-bond description.

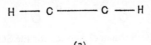

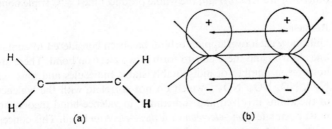

FIGURE 3.14. Valence-bond formulation of ethylene

In acetylene the basic σ skeleton is linear (Figure 3.15 (a)) and is described correctly if two linear sp hybrids are first constructed for each carbon. One of these brings about bonding to a hydrogen atom whilst the other overlaps with the sp hybrid on the other carbon.

H —— C ———— C —— H

(a)

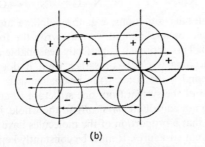

(b)

FIGURE 3.15. Valence-bond formulation of acetylene

There remain *two* unhybridized *p*-orbitals per carbon atom and a total of four electrons. Lateral overlap of both these orbital pairs gives rise to a *pair* of π-bonds which are at right-angles to each other (Figure 3.15 (*b*)) and the whole produces the $C\equiv C$ triple bond.

Resonance

Until now each overlapping orbital has been considered to contain one electron and these pair to form a two-electron bond. There are, however, a number of indisputably stable molecules and ions, e.g. N_2O, NO, NO_2, NO_3^-, which do not correlate with the 'valency' of the atoms involved and sometimes, in valence-bond theory, one must postulate a *one-electron* or a *three-electron* bond. This concept is one of the most awkward in the theory. The valence-bond way

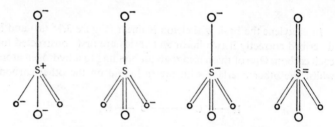

FIGURE 3.16. Resonance hybrids for the $SO_4^=$ anion

of explaining a species such as N_2O is to suggest *resonance* between different *canonical structures* or *resonance hybrids*, such as,

$$\overset{-}{N}=\overset{+}{N}=O \qquad \overset{+}{N}=\overset{-}{N}-O \qquad \overset{-}{N}=\overset{+}{N}=O$$

In other molecular situations, e.g. the sulphate anion, the same description can be applied and we may write the forms shown in (Figure 3.16). This approach is perfectly reasonable and valid providing its meaning is clear – or rather what it does *not* mean is clear. To exemplify the latter, (*i*) in resonance there is in no sense an oscillation of the system from one structure to another, (*ii*) the structures do not contribute *physically* to the whole, i.e. it must not be imagined that a proportion of the molecules have one form and the rest different structures. It must be constantly kept in mind that this is only a description of the molecular situation with which we are faced. The resonance hybrids may be regarded as contributors to the 'character' of the molecule. (*iii*) It is also wrong to suppose that the system stays in one canonical form for some fraction of the total time because this is tantamount to considering that there is an oscillation of the structure. Resonance merely *weights* the contributions of structures of one form or another to the real molecule.

The resonance approach is essential in valence-bond theory to

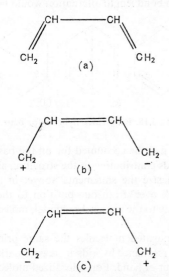

FIGURE 3.17. Resonance hybrids for butadiene

explain the delocalization of electrons which may prevail in some molecules. Good examples of this are furnished by unsaturated organic molecules like benzene and butadiene. Butadiene, C_4H_6, formally contains two C=C double bonds separated by a single C—C bond (Figure 3.17 (*a*)). Each half of the molecule could be described in the same way as was ethylene. Experimentally, it has been shown that the central C—C 'single' bond is shorter than in the paraffin analogue, *n*-butane. The implication is strong that the two unhybridized *p*-orbitals situated on either side of the central bond mutually interact and by this means a 'continuous π-bond' spreads from end to end of the molecule. In valence-bond-resonance terms the delocalization phenomenon is expressed by the set of resonance hybrids (e.g. Figure 3.17 (*b* & *c*)).

A second example is afforded by benzene. In benzene all the C—C bonds have equal length (1·397 Å), i.e. somewhere between the C—C (1·54 Å) and the C=C (1·35 Å) bond lengths. This would not be true if it were a cyclic compound composed of three linked ethylenes for which bond length alternation would be observed. The

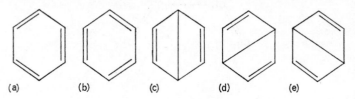

FIGURE 3.18. Resonance hybrids for benzene

experimental facts may be accounted for on the basis of a number of resonance hybrids contributing to the structure, as above (Figure 3.18). To re-emphasize the statements above, in (*a*) and (*b*) the bonds do not 'flick over' from one position to the next – all five structures contribute to the planar, hexagonal, molecule to a greater or lesser extent.

With larger conjugated molecules the same principles operate, but the number of resonance hybrids increases markedly and tends to get somewhat out of hand. For delocalized molecules in general, the molecular orbital approach to be developed in the next chapter is probably more satisfying.

PROBLEM FOR CHAPTER 3

1. Derive the general normalizing factor for the hybrid $(s + ap_z)$ if s and p_z are separately normalized.

CHAPTER 4

An Introduction to Molecular Orbital Theory

Diatomic Molecules

In Chapter 3 the valence-bond theory of molecules was outlined. This envisages orbital overlap as being of paramount importance in the formation of bonds between atoms. Hybridization was introduced to correlate the electronic structures of atoms with the observed symmetries of molecules and hybrid orbitals were constructed which could overlap with any other type of orbital. The equivalence of the hybrids explained the identity of bonds in highly symmetrical molecules.

We shall now adopt a second approach to molecular bonding which is based on the same principles as prevail for atoms, i.e. a number of orbitals are initially constructed ordered in an energy sequence. This process is carried out *without reference to electrons* in an exactly analogous way to that employed for atoms where there are *s-*, *p-* and *d-*orbitals, capable of accommodating specified numbers of electrons and which are filled up sequentially (the 'Aufbau Principle').

The approach to be developed here is called *molecular orbital* (m.o.) *theory*. The lines drawn between atoms to represent the bonds tend to get a little lost in this approach and later this particular feature may be turned to positive advantage. The quicker one is prepared to modify rigid 'valency' ideas the more rapidly can progress be made and, whereas valence-bond theory tends to perpetuate the emphasis laid on electron-pair bonding, in molecular orbital theory this is no longer done. No attempt is made to 'predict' the shape of a molecule as hybridization sometimes purports to achieve, but initially the experimentally known shape (or *symmetry*) of the situation is assumed. In that sense the m.o. approach is more realistic and leads to less confusion in general because it embodies fewer concepts open to misinterpretation on account of their quasi-physical nature. A well-known example to begin with is the hydrogen molecule. Let us first consider it as a pseudoatom (a cylindrical one) with two positive centres instead of only one. It can be regarded as a helium atom in which the two protons have been separated a little (we neglect the two neutrons as they are uncharged). This involves no conceptual difficulty and indeed we should anticipate a similarity between helium and the hydrogen molecule.

Now the two $1s$ electrons of helium move around the doubly charged nucleus and we may suppose that they would still do this if the two charges were separated.

The orbital in helium is an atomic orbital (Figure 4.1 (a)) containing two $1s$ electrons. In Figure 4.1 (b) when the positive charges are separated we call the system the hydrogen molecule though the

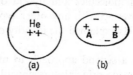

FIGURE 4.1. Relationship of helium and the hydrogen molecule

electrons may still be considered to be in an orbital which is analogous to the atomic orbital. Consequently such an orbital is named a *molecular orbital*. The initial concept is easy to grasp and may be extended readily to apply to a whole range of diatomic molecules. Later, this is what we shall attempt to do.

However, there are some more subtle points which should be looked into first. In helium each electron was located in a $1s$-orbital. Is this still true when the charges are separated? We should perhaps not expect an electron to be described by exactly the same orbital form although this should clearly be related in some way to a $1s$-orbital. Since there are now two centres of charge (two origins of coordinates) each electron may be considered to be described by an orbital centred on *both* protons A and B which 'goes over' to, or *correlates with*, a $1s$ atomic orbital containing the electron when the protons A and B coalesce to form helium. In order to obtain the form of the 'one-electron' molecular orbital it is clearly reasonable to use the atomic hydrogen orbitals $1s_A$ and $1s_B$ in some way. But how? No one knows the real answer to this question, but there is no lack of suggestions. The simplest and most obvious one is: if $1s_A$, $1s_B$ are appropriate to the separated protons, then let us try the new form $(1s_A + 1s_B)$ to describe an electron in the hydrogen molecule. Now the positive sign in the combination has no monopoly of correctness and so $(1s_A - 1s_B)$ must be considered also. These two forms are linear combinations of the $1s$ atomic orbitals centred on A and B. Such linear combinations have been met before (Chapter 3) when, e.g., an sp hybrid was constructed from separate s- and p-orbitals. At this point, care must be taken not to confuse the two situations; whereas in hybridization the s- and p-orbitals involved were centred on one and the same atom, the orbitals $1s_A$, $1s_B$ above are centred on two *different* hydrogen atoms.

The hybrid function $(s + p_z)$ is a single orbital and its partner is $(s - p_z)$ i.e. two atomic orbitals give rise to two hybrids. The orbital $(1s_A + 1s_B)$ is also a single orbital and the alternative combination $(1s_A - 1s_B)$ is its counterpart. Hence it should be noticed that the 'law of conservation of orbitals' still holds, i.e. from two original $1s$-orbitals we obtain *two* new linear combinations which are molecular orbitals.

In atoms the lowest energy orbital is called an s-orbital. Since we are seeking analogies between atoms and the hydrogen molecule (a cylindrical pseudoatom) a similar nomenclature may be used for molecular orbitals. Hence the molecular orbitals corresponding to s-orbitals are called σ-orbitals: σ is the first of a sequence of molecular orbital quantum numbers which correlate with the more familiar s, p, d, f atomic notation. The generic symbol for the molecular quantum numbers is λ and clearly π is the second in the series. For the molecular orbitals, as for atomic orbitals, Pauli's principle should operate so that our two molecular orbitals, $(1s_A \pm 1s_B)$ may accommodate, in all, four electrons. In hydrogen there are two electrons only and so it falls to us to decide which of the two molecular orbitals lies lowest in energy because both electrons will automatically occupy this one alone.

How can we decide between the two orbitals? It must first be realized that the forces which hold the two hydrogen atoms together in a bond do not differ under a molecular orbital description from those encountered in the valence-bond treatment, i.e. the net stabilizing forces are still essentially electrostatic attractions between an electron and the 'other' nucleus (described quantum mechanically). Similarly the destabilizing forces in the molecule are electron–electron and nucleus–nucleus repulsions. The difference in the two theories lies in how we *construct* the orbitals to describe the molecule.

The simplest v.b. wave function for the whole hydrogen molecule is,

$$\Psi_{\text{v.b.}} = \psi_A(1)\,\psi_B(2).$$

A comparable m.o. function, taking the positive sign throughout, is,

$$\Psi_{\text{mo.}} = (\psi_A + \psi_B)(1)\,(\psi_A + \psi_B)(2)$$

or, on expanding,

$$\Psi_{\text{m.o.}} = \psi_A(1)\,\psi_B(2) + \psi_B(1)\,\psi_A(2) + \psi_A(1)\,\psi_A(2) + \psi_B(1)\,\psi_B(2).$$

The first term is the same as the *complete* valence-bond function and the second duplicates the exchange term introduced as a second step in the valence-bond theory (Chapter 3). The remaining *two* terms will be recognized as those representing *ionic* contributions to the bond. Hence the simplest m.o. wave function for the hydrogen molecule incorporates most of the advanced features of the valence-bond function. However, harking back to eqn. (3.5) in Chapter 3 which described a bond thus

$$\Psi_{\text{total}} = \psi_{\text{covalent}} + a\psi_{\text{ionic}}$$

this equation recognizes that the contribution of ionic terms (governed by 'a') to a bond will not, in general, be the same as that of the covalent terms. Indeed, in a homopolar case, such as H_2, the former will be very small. The simple m.o. wave function for hydrogen is therefore deficient on this score because the 'a' factor in front of the ionic terms is unity and hence their contribution is much overestimated.

The important point which emerges from these deliberations is that the m.o. with the *positive* sign corresponds to the lower energy of the two and is, in addition, lower in energy than an atomic orbital centred on a single hydrogen atom. Because of this the orbital is said to be a *bonding* orbital. On the other hand if we carry out the analysis with the linear combination having the *negative* sign a level *greater* in energy than that of the isolated hydrogen atom results. Hence this type of orbital is called an *antibonding* orbital (designated by *).

There is a second instructive way in which this may be considered. The antibonding molecular orbital changes sign between the atoms and so the wave function must pass through zero halfway between them. Now in atoms the value of ψ^2 at a point was equated to the electron density at that point and so applying the same reasoning here the electron density is zero halfway between the nuclei. Thus the electrons are not concentrated between the nuclei and are 'out of position' for holding the atoms together. For the m.o. with the positive sign, however, the electron density is finite at all points between the atoms and it is clear why the latter describes a bonding situation whereas the former is antibonding. This correlates well with the conclusion of the argument based on energy.

Homopolar Diatomic Molecular Systems with More than Two Electrons

In this development we have achieved something rather more far-reaching than just an explanation of the bonding of two hydrogen atoms. What we have really done is to construct the first two orbitals which lie in a molecular orbital energy sequence just as was derived from atoms. Now for a hydrogen-like *atom* the first two orbitals are in the energy order $1s < 2s$ and if electrons are fed into these, as the Aufbau Principle dictates, we have the configuration H $1s^1$; He $1s^2$; Li $1s^2 2s^1$; Be $1s^2 2s^2$, i.e. four elements possessing from one to four electrons. An identical procedure may be carried through using the two *molecular* energy levels that have been constructed. The first four species are,

$$H_2^+(1\sigma)^1; \quad H_2(1\sigma)^2; \quad H_2^-(1\sigma)^2(1\sigma^*)^1$$
$$\text{(or } He_2^+(1\sigma)^2(1\sigma^*)^1); \quad He^2(1\sigma)^2(1\sigma^*)^2.$$

There are several features here worthy of consideration and they give rise to both surprise and satisfaction. Firstly, let us consider the ions H_2^+, H_2^- and He_2^+. The first is just two hydrogen atoms held together by one bonding electron and so is a stable unit with a bond energy approximately one-half that of H_2. Although we do not get it in test-tubes in the laboratory, it has been detected spectroscopically. In valence-bond jargon this would be an example of a one-electron bond (something to become puzzled over!) but here we have obtained it quite naturally and it has given us no more trouble than did the hydrogen molecule with its two electrons. The latter two species have the same electronic structure, but will differ energetically in other ways into which we need not go here. We may regard

H_2^- as incorporating a three-electron 'bond' without any dismay. Here, however, the situation differs from H_2 or H_2^+ in that one of the electrons is antibonding and this 'cancels out' a bonding one; H_2^- is then held together, effectively, by one electron. Perhaps the most surprising is He_2^+; we might think that there are no entities in which a helium atom was bonded to another, but here is an ion held together by one nett bonding electron – weakly to be sure – but still bonding.

The lowest energy orbital was referred to as the *σ-bonding* orbital and so the other, $(1s_A - 1s_B)$, is called a *σ-antibonding* orbital.

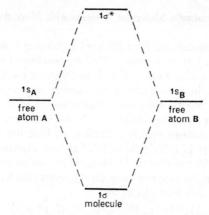

FIGURE 4.2. Correlation diagram for $1s$-orbitals

Because the former is of lower energy than the isolated hydrogen atoms it holds them together in the form of a molecule whereas the latter, which has energy greater than the original pair of atoms, is a repulsive state, i.e. it holds the atoms apart. Hence if the antibonding level of hydrogen contained the two electrons instead of the bonding one then the molecule would fall apart. However, in the ground state of the molecule the antibonding orbital contains no electrons; both of these occupy the bonding orbital. The meanings of the terms bonding and antibonding are obviously of crucial importance to an understanding of the energetics of molecule formation.

The situation may be further illustrated by a *correlation diagram* (Figure 4.2) which shows the way in which two *s*-atomic orbitals on separated atoms come together and give rise to the σ and σ^* levels. Notice that, for a homopolar system like H_2, the antibonding level is about the same distance above the energy zero (an obvious choice of this is the energy of an electron on an isolated H atom) as the bonding level falls below it.

For He_2 there are four electrons in the structure, two bonding and two antibonding, and the latter completely cancel the former leading to no resultant bonding: on this basis we predict that He_2 will not be formed. This is indeed a satisfying conclusion because the molecule does not exist and so by our simple approach experimental observation has been confirmed. It is worth while at this juncture to re-emphasize an initial point: the molecular orbitals were made up first of all without any regard to electrons and these were only later fed in when the energy sequence of the orbitals was known. This is our basic principle.

Let us now move on to consider the $2s$-orbitals. As before, the $2s$-orbitals on two distinct nuclear centres may be taken together in *symmetric* (with $+$ sign) and *antisymmetric* (with $-$ sign) linear combination thus: $(2s_A + 2s_B)$, $(2s_A - 2s_B)$ and, again as before, the first has the lower energy of the two. A new point now arises; because it is formed from $2s$ atomic orbitals the symmetric combination has *higher* energy than *either* the previous 1σ or $1\sigma^*$ combinations and the antibonding $2\sigma^*$ orbital lies yet higher. The modified correlation diagram is shown in Figure 4.3.

There is now room for four more electrons and so we have

$$Li_2^+ \quad He_2^-(1\sigma)^2(1\sigma^*)^2(2\sigma)^1 \quad Li_2(1\sigma)^2(1\sigma^*)^2(2\sigma)^2$$
$$Li_2^- \quad Be_2^+(1\sigma)^2(1\sigma^*)^2(2\sigma)^2(2\sigma^*)^1$$
$$Be_2(1\sigma)^2(1\sigma^*)^2(2\sigma)^2(2\sigma^*)^2.$$

These may be examined in turn and discussed as before for the 1–4 electron cases. The interest is somewhat diminished however because, with the possible exception of Li_2, none of them is important chemically.

Further conclusions, of much greater importance, may be reached by moving on to a consideration of the $2p$-orbital combinations. Boron has the electronic configuration $1s^2 2s^2 2p^1$ and so we shall now have to investigate the properties of molecular orbitals formed from linear combinations of $2p$-orbitals. If there are two atomic

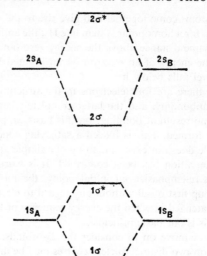

FIGURE 4.3. Correlation diagram for $1s$- and $2s$-orbitals

centres, on each of which is a single $2p$-orbital and these are directed towards each other as in Figure 4.4 then two linear combinations of these, symmetric and antisymmetric may be made up as before. Once again the orbital with the positive sign has the lowest energy which, moreover, is higher than that of the $2\sigma^*$-antibonding orbital just as the 2σ was higher than the $1\sigma^*$-orbital. Clearly the lower bonding level ($+$ sign) formed from the $2p$ atomic orbitals can accommodate two electrons and is directed along the line of centres between the atoms. Because of the latter property the orbital is still

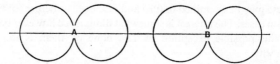

FIGURE 4.4. Orbitals giving rise to $2p_\sigma$ molecular levels

called a σ-orbital ($2p_\sigma$), *despite* its being formed from $2p$ atomic orbitals, and it has cylindrical symmetry along the interatomic axis. In fact, we can now generalize and state that any bond which has cylindrical symmetry along the line of centres is a σ-bond independently of which type of atomic orbitals form it.

With these new $2p_\sigma^-$ and $2p_\sigma^*$-orbitals to be used there is now room for four more electrons in the energy sequence and we can re-apply the Aufbau technique. Hence the bonding pattern for the molecule B_2 will incorporate two electrons in the $2p_\sigma$ level, the total molecular configuration being

$$B_2(1\sigma)^2(1\sigma^*)^2(2\sigma)^2(2\sigma^*)^2(2p_\sigma)^2.$$

A problem now arises that must be resolved before any further progress can be made: we have considered only one $2p$-orbital on each atom. What is to be done about the other two which are

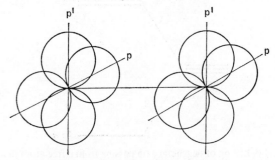

FIGURE 4.5. Orientation of $2p$-orbitals for π-bonding

orientated at right-angles to the line of centres and are, in addition, mutually perpendicular? Evidently these two orbitals on each atom are equivalent to each other in all respects because one pair (p'_A, p'_B in Figure 4.5) may be turned into the other (p_A, p_B) on both atoms simply by rotating them by 90° around the molecular axis. However, there is no *single* operation such as this which will turn either of the pairs into the two p-orbitals directed along the line of centres. We conclude that the two p-orbital pairs normal to the σ-bond differ from the pair forming it but are equivalent to each other. If we now hark back quickly to the atomic situation we find an immediate analogy: all the three p-orbitals on an atom are equivalent to each other in all respects – they have the same energy, and constitute a triply degenerate set. It follows that the mutually perpendicular pair in the linear diatomic molecule are doubly degenerate and have the *same* energy but the third, which forms the σ-bond, has a *different* energy. The situation thus changes drastically on making the transition from the free atom to the linear molecule (Figure 4.6). The degeneracy of the free atom orbitals is reduced from three to two and is said to have been *lifted* when the linear molecule is formed. Such lifting almost invariably takes place when a transition is made from an atomic to a molecular situation.

But to return to the initial problem. The next stage is clear; the two pairs of orbitals at right-angles to the σ-bond can obviously be taken together in linear combinations just as for p_σ-orbitals and

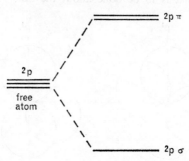

FIGURE 4.6. Lifting of degeneracy on passing from a free atom to a linear molecule

from them will result four molecular orbitals, $(2p_A + 2p_B)$, $(2p_A - 2p_B)$, $(2p'_A + 2p'_B)$ and $(2p'_A - 2p'_B)$. Furthermore the two m.o.'s with the + signs have the same energy and so also have the two antisymmetric orbitals. Since they *all* originate from $2p$ atomic orbitals, just as did the $2p_\sigma$ pair, their absolute energies are similar to the latter, although not identical.

A final level scheme for the molecular orbitals formed from all combinations of $2p$ atomic levels may now be drawn up (Figure 4.7). The doubly degenerate pair of bonding molecular orbitals can accommodate four electrons. This type of molecular orbital which is formed from orbitals not directed along the line of centres is called a π-orbital and has molecular orbital quantum number, $\lambda = 1$. For a homopolar diatomic molecule therefore a *pair* of

π-molecular orbitals, or π-*bonds*, may exist which are mutually perpendicular and also normal to the σ-bond. The complete correlation diagram is now shown in Figure 4.8 and the whole set of levels will take, in all, twenty electrons.

We have now generated a complete set of molecular orbitals falling in an established energy sequence and the electrons can be fed in one by one to obtain the electronic structure of any desired

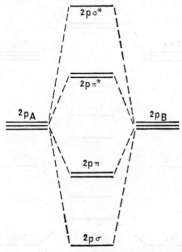

FIGURE 4.7. Correlation diagram for $2p$-orbitals

diatomic molecule having up to twenty electrons. The next homopolar molecule of interest, following B_2, is nitrogen. Each nitrogen atom of the pair possesses seven electrons and all that needs to be done is to place the whole fourteen into the complete level scheme. Obviously they will fully occupy the lowest seven levels which leads to the molecular configuration,

$$(1\sigma)^2(1\sigma^*)^2(2\sigma)^2(2\sigma^*)^2(2p_\sigma)^2(2p_\pi)^2(2p_\pi)^2.$$

The last four electrons have the same energy.

What does all this mean in chemical terms? The first eight electrons constitute two bonding–antibonding pairs and no nett bonding

H

can result from these. The last three levels, however, are completely different. There are no antibonding electrons opposed to these at all and so *all six* hold the two nitrogen atoms together. The three pairs of electrons thus form a triple bond between the two atoms, which readily accounts for the great chemical stability and heat of dissociation of the molecule. Furthermore the bonding scheme tells us that two of these electrons are of σ-bonding type while the other four

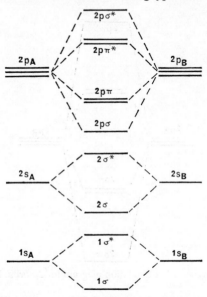

FIGURE 4.8. Complete correlation diagram for 1s-, 2s- and 2p-orbitals

are in two π-bonds which have the same energy. We can also readily predict the effect of ionization on the bond dissociation energy of the system. Removal of one electron, giving N_2^+ and leaving only five bonding electrons, *weakens* the N—N bond and, alternatively, if one is added (N_2^-) this occupies an antibonding orbital and so will have precisely the same effect.

The next case of interest is the oxygen molecule. Here there is a total of sixteen electrons and its electronic structure is obtained by adding two further electrons to the nitrogen configuration. The next energy levels in the scheme (Figure 4.8) are the doubly degenerate antibonding $2p_\pi^*$ pair and hence the two electrons are fed in here. Now the lowest energy situation (in the absence of external forces) prevails when electrons are *unpaired* and have *parallel* spins (a rule stemming from interelectronic repulsions). Consequently the 'last' two electrons are placed one in each of the $2p_\pi^*$ levels and the electronic structure of O_2 is (Nitrogen core) $(2p_\pi^*)^1(2p_\pi^*)^1$.

Several properties of oxygen may now be predicted:

(1) There is one *nett* double bond (one σ, one π) between the atoms, because two of the π-bonding electrons are 'cancelled out'.
(2) The two unpaired electrons in the π-antibonding orbitals will confer the property of *paramagnetism* on the molecule.
(3) Because it differs from nitrogen in its possession of two antibonding electrons, its bond dissociation energy will be *lower* than that of N_2.
(4) Ionization to O_2^+ involves the loss of an *antibonding* electron and so *strengthens* the O—O bonding.

It should be noted that, were the valence-bond approach to the bonding in the molecule to be adopted, then we would consider overlap between the two $2p$-orbitals directed along the molecular axis (σ-bond) and a second pair at right-angles to it (π-bond). This, at first sight, appears to be what we have suggested from molecular orbital theory also. The difference lies in that the valence-bond treatment pairs all the electrons on overlap whereas two unpaired ones manifest themselves from the molecular orbital approach. Experimental examination of the magnetic properties of oxygen has shown that the latter is correct.

For fluorine we add two more electrons to the level pattern (Figure 4.8) which fill in the two vacant places in the antibonding $(2p_\pi^*)$ levels and hence 'neutralize' all four π-bonding electrons which were so important in stabilizing nitrogen. This shows that the fluorine molecule is bonded by a single σ-bond, a fact which corresponds with its known low heat of formation. Finally, on adding two further electrons (Ne_2), all orbital vacancies are filled and all bonding electrons matched by antibonding ones. The Ne_2 molecule has therefore no nett bonding energy and should not be formed. As is well known, it has not yet been prepared.

Heteropolar Diatomic Molecules

It is clear from the foregoing account that the molecular orbital approach yields a very satisfying picture of these diatomic molecules which correlates, at this level, semi-quantitatively with experimental observations. However, in order to get the full benefit of molecular orbital theory we should look at a typical heteropolar case – nitric oxide is eminently suitable. Using valence-bond theory we must write a set of resonance hybrids for this molecule and these make it appear 'different' from others. Under the present scheme, however, a simple description of it may be gained by setting up a new correlation diagram. This will be slightly different from Figure 4.8 in that the atomic orbitals on each side of the diagram will be positioned at different points on the absolute energy scale. This is because s- and p-orbitals on different atoms do not have equal initial energy on account of the difference in the atomic nuclear charges. Nevertheless the ordering of the *molecular* energy levels is precisely the same (Figure 4.9).

On filling up the orbitals, two by two, in the usual way, we find that nitric oxide has one antibonding electron, i.e. it has essentially the same electronic structure as N_2^-. If this odd electron is removed, giving NO^+, the N—O bond should become stronger and there is spectroscopic evidence that it does. In fact the NO^+ cation is a well-known chemical species present in some coordination complexes.

It is interesting that, using a level scheme such as proposed above for a *heteropolar* diatomic molecule, we would also predict the existence of the molecule NeF. This would be isoelectronic with Ne_2^+ and would be held together by one σ-bonding electron overall.

What is perhaps most significant is that all these features arise quite naturally from our original premises and there is no need to postulate correct but easily misinterpreted 'phenomena' like resonance or hybridization. The latter does not enter into pure molecular orbital theory at all. There should be little difficulty with the concepts of molecular orbital theory if it is remembered that they are intimately related to those of atoms. The idea of 'sticks between atoms' representing bonds has been almost completely lost and there is no need to conjecture how an odd electron (as in NO) would be represented in terms of 'sticks'. In fact, since this particular electron is antibonding in nature we would be hard put to it to find an analogy in physical terms – in the author's view such an analogy is not needed. Lines drawn between atoms are useful in order to clarify geometry, but need have no further meaning.

One of the outstanding differences between the m.o. approach and the previous valence-bond treatment is worth re-emphasizing. This lies in the treatment of electrons by each. In the former these

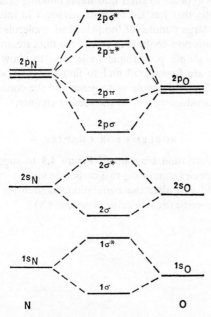

FIGURE 4.9. Complete correlation diagram for nitric oxide

are first 'stripped off'; we then linearly combine the *atomic orbitals* on the component atoms and build up an energy-level series. Finally the electrons which were stripped off are fed two by two (as far as possible) into the energy levels. It should be noticed particularly that there is no need to 'unpair' any of them initially, to 'promote' them, or to 'hybridize' the orbitals. The 'difficulty' met in Chapter 3 regarding the expected 'zero valency' of beryllium on the basis of its ground-state configuration $(1s^2 2s^2)$ disappears because we simply remove the electrons considered to be 'available' and feed them back into the m.o's of the new system we are describing (the molecule). It should be constantly borne in mind that in the m.o. approach the electrons, in principle, encompass and contribute to the stability (or otherwise!) of the whole molecule.

It is the realization that two atoms do not need a pair of electrons between them in order to enter into stable bonding (although this is often desirable) that has led to most advances in interpreting the bonding in a large number of 'non-classical' molecules.

What remains now to do is to extend the m.o. treatment in order to deal with simple polyatomic molecules, to show how linear combinations are formulated and to find the rules which govern their forms. This will enable us to generalize the concepts put forward in an introductory way in the present chapter.

PROBLEMS FOR CHAPTER 4

1. Use the correlation diagram of Figure 4.8 to suggest possible chemical species containing two carbon atoms.
2. How would you alter the correlation diagram for nitrogen in order to describe the isoelectronic species CO?

CHAPTER 5

Molecular Symmetry and Molecular Orbital Theory for Polyatomic Molecules

Previous chapters have dealt with atomic orbitals and the ways in which they may overlap and this leads naturally to an explanation of chemical bonding and the concepts of σ- and π-bonds. It was also shown qualitatively how the resultant overlap between two atomic orbitals may be zero, i.e. when the two orbitals are orthogonal to each other. What we now finally wish to do is to develop and apply to the bonding problem in polyatomic molecules certain elegant and extremely powerful notions which are a consequence of *molecular symmetry*.

Symmetry

All objects possess some degree of symmetry and, equally clearly, some have more than others. The amount of symmetry an object has may be roughly related to the number of ways in which it may be turned yet left indistinguishable from its initial form, e.g. this printed page will only look the same if we either leave it alone or turn it through 360° whereas a billiard ball looks the same whatever angle it is rotated through. These examples are extremes in terms of symmetry and most objects lie somewhere in between.

It is useful, and possible, to make the idea more quantitative and we may assess the symmetry of an object in terms of the *number of operations* which may be performed on it so as to leave it looking the same. These are a consequence of its intrinsic *symmetry elements*. For example, consider the following objects (Figure 5.1 (*a–l*)).

What symmetry operations may be performed on the letter A? If it is rotated around its apex by 180° out of the plane of the paper it will be unchanged; similarly if we imagine two mirrors, one in the plane of the paper and another vertically through the apex at right-angles to the first, then reflection in either of these will also leave the figure looking the same. Obviously if we do nothing (or rotate the letter by 360°) it will also be unchanged. In fact it turns out that these are the only symmetry operations which can be carried out on the letter A and they comprise two rotations (one is by 360°) and two reflections. This set of operations is very important and we shall

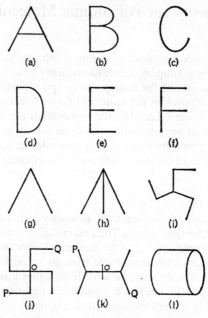

FIGURE 5.1. Symmetry in geometrical figures

refer to it time and again in this chapter. Moving on to the letter E it may soon be established that precisely the same set of operations that described A exist for this too, albeit the 180° rotation axis is a horizontal and not a vertical one.

Thus, despite the apparently different shape of E, it is of the *same symmetry type* as is A and it is instructive to think up other articles which would have the same sort of symmetry. For object 5.1 (*i*) the symmetry operations are (*i*) leave it alone, (*ii*) rotate it in the plane by 120° (in either direction), (*iii*) reflect it in the plane of the paper. Thus it differs from A or E in being specified by a different *group of operations*. Consider next 5.1 (*h*) which is a three-pronged umbrella. This we could rotate by 120° in either direction or reflect it vertically across the central line which is a mirror plane (clearly there are three of these: one through each prong). Each of these operations would leave the object unaltered. Notice that this group of operations is related, but not identical, to that of object 5.1 (*i*) because they both possess the 120° rotation axis.

All the elements of symmetry of the remainder of the figures may be discovered. Among these operations there will occur one which is called the inversion operation. This may be illustrated by reference to 5.1 (*j*): any point on the figure has an equivalent point the other side of the centre O, i.e. in the vertically opposite direction and at the same distance from O, e.g. the points Q and P are related to each other by inversion. The centre O is called the *centre of symmetry* of the object. This element is possessed by Figures 5.1 (*j*), 5.1 (*k*), 5.1 (*l*) but is not present in all the rest. In a system of Cartesian coordinate axes the inversion operation is the same as taking the point $P(x, y, z)$ and transforming it to the point $Q(-x, -y, -z)$. This is apparent from 5.1 (*j*) if O is taken as the origin, although 5.1 (*j*) has no extension in the z direction which is perpendicular to the plane of the paper.

It transpires that *all* objects are described by one or other of 32 groups of operations, and so in this simple way we have produced a system of classification for all objects. It now becomes relevant to apply this approach to some simple mathematical functions in order to find what their symmetry properties are.

Even and Odd Functions

Perhaps the simplest functions that are encountered in mathematics are, $x, x^2, x^3, \ldots, x^n$ and the trigonometrical ones $\sin x$ and $\cos x$. Their graphs, for positive and negative x are shown overleaf.

A striking difference between x^2 and $\cos x$ on the one hand and x, x^3 and $\sin x$ on the other should be immediately apparent. For the former pair the value of the function at any positive x is the same as it is for equal negative x whereas for the latter group the magnitude of the function is the same but its *sign* is reversed. We say that a

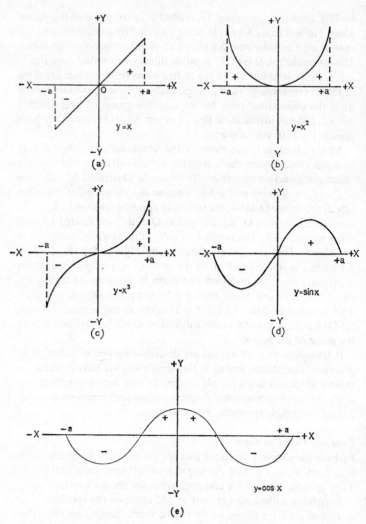

FIGURE 5.2. *Even* and *odd* functions

function is *even* if it does not change sign on substituting $-x$ for x but it is *odd* if it does. It follows readily that all *even* powers of x are *even* functions whereas all *odd* powers of x are *odd*. Moreover $\sin nx$ is always odd and $\cos nx$ always even.

It should now be recalled that changing x for $-x$ constitutes in one dimension the symmetry operation that was called *inversion* and, in this section, we have applied it, not to a geometrical figure, but to a mathematical function. This shows that symmetry operations can be applied to functions in the same way as if they were objects.

Integrals of Odd and Even Functions

Odd functions possess an important property. If an *odd* function is integrated between the constant limits $(a, -a)$ *the result is always zero*. Thus

$$\int_{-a}^{a} x \, \mathrm{d}x = 0, \qquad \int_{-a}^{a} x^3 \, \mathrm{d}x = 0, \qquad \int_{-a}^{a} \sin x \, \mathrm{d}x = 0.$$

This can be readily understood from Figure 5.2 because a definite integral measures the area under a graph and *odd* functions always have the same amount of 'negative area' as 'positive area'; and these two mutually cancel on addition by integration. However, if we integrate an *even* function of x between the same limits *the result is not zero*. The *actual* value of such an integral is unimportant; it suffices to know that such an integral has a non-zero value. Now the above are 'single' functions; what happens if the integral is a *product* of two or more functions, e.g. $\int_{-a}^{a} x \cos x \, \mathrm{d}x$? The question is readily answered by first writing x^2 as $x.x$, x^3 as $x.x^2$ and x^4 as $x^2.x^2$.

Thus
$$\int_{-a}^{a} x^2 \, \mathrm{d}x = \int_{-a}^{a} (x.x) \, \mathrm{d}x \neq 0$$
$$\qquad\qquad\qquad odd \times odd$$

$$\int_{-a}^{a} x^3 \, \mathrm{d}x = \int_{-a}^{a} (x.x^2) \, \mathrm{d}x = 0$$
$$\qquad\qquad\qquad odd \times even$$

and
$$\int_{-a}^{a} x^4 \, dx = \int_{-a}^{a} \underset{even \, \times \, even}{(x^2 . x^2)} \, dx \neq 0$$

From these cases arises a general rule: for functions of one variable (x, say) *the integral of any function which has, overall, odd symmetry is zero while that of an overall even function is non-zero.* The following equalities help in using the rule,

$$odd \text{ function} \times odd \text{ function} = even \text{ function}$$
$$even \text{ function} \times odd \text{ function} = odd \text{ function}$$
$$even \text{ function} \times even \text{ function} = even \text{ function.}$$

The primary question may now be answered fully and we have

$$\int_{-a}^{a} x \sin x \, dx \neq 0 \qquad \int_{-a}^{a} x \cos x \, dx = 0, \qquad \int_{-a}^{a} x^2 \cos x \, dx \neq 0$$

and so on. Obviously it can readily be predicted whether or not an integrand which has symmetry properties will vanish without actually performing the integrations. This important conclusion, which we have shown to hold for functions of one variable, is also true for three dimensions and it has immense influence on chemical and spectroscopic theory. Let us now apply it to two of the concepts introduced in Chapters 3 and 4, namely orbital overlap and linear combinations of atomic orbitals.

Orbital Overlap

The magnitude of the overlap between two orbitals is governed by an overlap integral which has the form $\int \psi_a \psi_b \, d\tau$. Here ψ_a and ψ_b are the two overlapping orbitals and the integral is bounded in x, y, z by $-\infty$ to $+\infty$. It should be noticed that this is identical in form to the simple integrals we have been considering above and hence it should be amenable to the same sort of arguments as used there. Now $\int_{-a}^{a} x \cos x \, dx$ is zero because x is *odd* but $\cos x$ is *even*, i.e. they are not of the *same symmetry type*. It is reasonable to expect this rule to apply to the overlap integrals also, i.e. the integral will be zero if ψ_a and ψ_b are not of the same symmetry type. The fact that ψ_a and ψ_b, are not, in general, simple one-variable functions like x or $\cos x$, causes no breakdown of the rule. The concept can also be studied pictorially and Figure 5.3 illustrates some important cases.

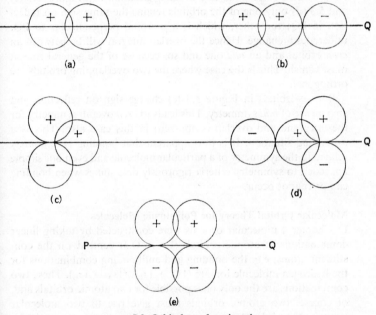

FIGURE 5.3. Orbital overlap situations

Now functions of x are *even* and *odd* with respect to change of x for $-x$, i.e. the inversion operation. But, as we have seen, there are other possible symmetry operations such as rotations and reflections, e.g. in Figure 5.3 PQ is a mirror plane of symmetry which gives rise to a reflection operation. What happens to the signs of the orbitals when they are reflected in it? In (a) the positive signs of two s-orbitals clearly do not change on reflection in the plane: in (b) also, because the p-orbital is lying in the mirror plane, the same signs are again retained on reflection though they are not all positive in this case (a p-orbital has one negative lobe). Retention of sign in this way means that the orbitals of Figures 5.3 (a) and 5.3 (b) are *even* to the reflection operation. From previous arguments, if *both* the orbitals in the overlap integral have *even* symmetry then the integral will be non-zero. Thus there is nett orbital overlap in Figures 5.3 (a) and 5.3 (b) which correspond to σ-bonding situations. In Figures 5.3 (c) and 5.3 (d) only one of the orbitals retains the same sign on reflection in PQ; the other, which is a p-orbital perpendicular to PQ, undergoes a change. Hence the overlap integral will be between an *even* orbital and an *odd* one and so, because of the general rule, it must vanish. This is the case where the two overlapping orbitals are orthogonal.

Both p-orbitals in Figure 5.3 (e) change sign on reflection and so *both* are of *odd* symmetry. This leads to *even* overall symmetry for the integrand and overlap is non-zero in this case also. This is a π-bonding situation. It is thus apparent that bonding is intimately related to the symmetry of a particular molecule and even our simple approach to symmetry criteria rigorously determines when bonding can or cannot occur.

Molecular Orbital Theory for Polyatomic Molecules

In Chapter 4 molecular orbitals were constructed by taking linear combinations of atomic orbitals (L.C.A.O's) centred on the constituent atoms, e.g. the bonding and antibonding combinations for the hydrogen molecule ion are $(1s_A + 1s_B)$, $(1s_A - 1s_B)$. These two combinations are the only ones possible for two atomic orbitals and, of course, two atomic orbitals must give rise to two molecular orbitals. Now it has been suggested that all molecules (not only the diatomics of Chapter 4) should be described by molecular orbitals in turn built from linear combinations of atomic orbitals. In the valence-bond approach a large covalent molecule was described bond by bond in terms of electron pairs essentially localized between juxtaposed atoms; by contrast in the m.o. approach electrons are described by orbitals which encompass the whole molecule which suggests that the electrons are delocalized over all of it. Thus the two electrons of H_2 circulate around both nuclei. This physical picture correlates with wave functions which include terms representing the contribution of covalent, exchange and ionic terms to the whole. The 'law of conservation of orbitals' must also operate for polyatomic molecules so that if the system contains 100 atoms, each contributing one orbital, then 100 m.o's can be constructed from these. The whole aggregate of 100 orbitals would then *together* describe the molecule. There are, however, certain restrictions placed on the

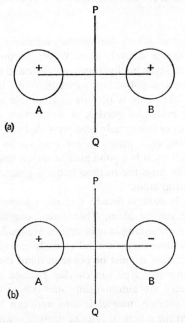

FIGURE 5.4. Symmetry of linear combinations

m.o's; they cannot be chosen completely arbitrarily and the rules which determine how they may be constructed must be formulated. Anticipating a little, it is the *symmetry* of the molecular situation which determines rigorously the form of the m.o's.

The first step is to study the two molecular orbitals of H_2 and to discover their orthogonality properties. They are shown in Figure 5.4. The two orbitals (bearing in mind that a linear combination is a *single* orbital) differ in their behaviour to reflection across the mirror plane PQ because the first is *symmetric* (or *even*) whereas the second is *antisymmetric* (or *odd*) to this element. Hence we know that the two molecular orbitals must be of different symmetry type and because of this are necessarily mutually orthogonal. It may therefore be concluded that precisely the same rules govern the overlap of molecular orbitals as do atomic orbitals. Moreover this basic rule is quite general for all molecules however complex.

Water

There is one feature which distinguishes a certain group of molecules; they contain a 'central' atom. Water, boron trifluoride and methane are of this type but hydrogen, ethylene and benzene are not. The most appropriate polyatomic molecule with which to open the discussion is water. This is because it is simple to treat and will provide all the principles needed. A description of water under valence-bond theory has already been presented and we want now to redescribe it using m.o. theory in order to see in what ways the interpretation differs. It is a good plan to review first the procedure to be followed because the method is quite general for molecules possessing a central atom.

Briefly, what is done is first to find the symmetry types of the orbitals on the central atom. Then linear combinations of the orbitals on the surrounding atomic centres are made up and finally overlapped with those on the central atom. In carrying out the final stage of the operation it must be borne in mind that only orbitals of the same symmetry type can overlap because two orbitals of different symmetry are automatically mutually orthogonal. Consequently their overlap integral is zero and can bring about no bonding between the atoms. It is quite straightforward to carry out the whole procedure to investigate the bonding in water. This will be done in detail and should be studied with care.

The electronic structure of the oxygen atom is $1s^2 2s^2 2p^4$. The $1s$ electrons of the oxygen do not enter into the bonding and are left atomic. This is, of course, an approximation but it is a justifiable one as we may check by referring back to Figure 4.8 of Chapter 4. This shows that *both* the 1σ and $1\sigma^*$ orbitals are filled in the oxygen molecule and so the $1s$ electrons of oxygen could have been left out of consideration in the first place, i.e. they could have been left atomic.

The bonding diagram would then not have featured the 1σ and $1\sigma^*$ orbitals at all. In the 'heavier' first-row elements such as oxygen it is proposed always to leave the $1s$ electrons atomic when considering bonding.

What are the symmetry properties of the atomic orbitals centred on the oxygen atom? Figure 5.5 shows how they are situated and, so that the disposition of the oxygen orbitals may be fully appreciated, the $H\widehat{O}H$ bond angle has been artificially lessened. It is only the

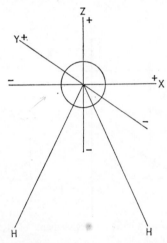

FIGURE 5.5. Orientation of oxygen orbitals in H_2O

symmetry properties of *angular functions* which are important. The *radial function* is spherically symmetrical, is unaffected by symmetry operations, and need not be considered further. To avoid complicating Figure 5.5 overmuch the lobes of the p-orbitals have not been drawn in but have been represented by positive and negative signs along the x-, y- and z-axes, i.e. where the p_x-, p_y-, p_z-orbitals themselves have positive and negative regions. Reference back to Chapter 2 will be useful here.

Symmetry principles were developed at the beginning of this chapter using physical objects but it was also shown that angular functions could be manipulated rather as though they were objects and symmetry operations could be performed on them. What symmetry operations can be performed here? Firstly we can only carry out those operations made possible by the symmetry elements possessed by the water molecule. This molecule is shaped like the letter 'A' (minus the horizontal bar) and hence possesses the same symmetry elements as this letter: these allow the same set of operations to be performed (Figure 5.6). These are (*i*) leave-it-alone, (*ii*) rotate around the vertical axis (*z*), (*iii*) reflect across the mirror plane (I), (*iv*) reflect across the mirror plane (II). These operations

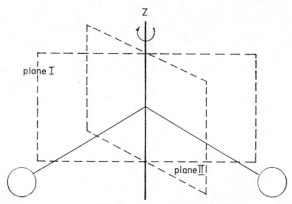

FIGURE 5.6. Symmetry elements of H_2O

should be committed to memory as they will be extensively used throughout this chapter. They are the only operations that can be carried out on an angular figure such as this.

Now when we operate on an orbital we shall find that it will either remain looking the same or the negative and positive signs on the axes will be reversed. An orbital which is unaltered by an operation could be regarded as being multiplied by the factor, $+1$, whereas if its positive and negative signs are reversed, it is multiplied by -1. Use of models will be found to help considerably in carrying out this procedure. It is convenient to collect the results of performing the operations together in a table (Table 5.1). Across the top of the table are written the operations, and down the rows, the orbitals under consideration. The effect of the operations will then be written as $+1$ or -1 in the table.

First the oxygen $2s$-orbital. If we 'leave it alone' it will obviously be unchanged and so we write $+1$ in the table for this operation. Now if the $2s$-orbital is rotated by $180°$, it is again unaltered so under '$R_{180°}$' we also write $+1$. Reflection in the plane of the paper does not change the orbital in any way either so we write $+1$ under 'Ref. (I)', and, finally, reflection in a plane perpendicular to the plane of the paper again leaves it unaltered. The complete symmetry of the $2s$-orbital to these operations is thus specified precisely by a

Table 5.1 *Effect of operations on orbitals of oxygen*

	'leave-it-alone' i.e. identity	Rotation by $180°$ ($R_{180°}$)	Reflection across I (Ref. I)	Reflection across II (Ref. II)	Symbol for represent- ation
$2s$	$+1$	$+1$	$+1$	$+1$	a_1
$2p_z$	$+1$	$+1$	$+1$	$+1$	a_1
$2p_x$	$+1$	-1	$+1$	-1	b_1
$2p_y$	$+1$	-1	-1	$+1$	b_2
	$+1$	$+1$	-1	-1	a_2

set of four numbers all of which are $+1$. The set of numbers could therefore be said to *represent* the $2s$-orbital and they are called a *representation* of the $2s$-orbital. Each separate number *characterizes* the behaviour of the orbital under a particular operation – so it is called the *character* of that operation.

Let us now study the other orbitals. The p_z-orbital lies along the vertical z-axis. The positive and negative signs illustrate its functional form. Applying the four operations we get:

(a) leave-it-alone – it remains unchanged and therefore has
character $+1$.

(b) $R_{180°}$ -do- $+1$.

(c) Ref. (I)   -do- $+1$.

(d) Ref. (II)   -do- $+1$.

The effect of operations (b), (c) and (d) on this orbital follows automatically from the fact that this p-orbital lies along the z-axis and also on the intersection of the two reflection planes. (Operation (*a*), the 'leave-it-alone' operation, always has the same effect independent of which orbital is studied and hence is called the *identity* operation.) The set of characters for $2p_z$ are recorded in Table 5.1.

Consider now the $2p_x$-orbital: this lies along the x-axis and at right angles to the $2p_z$-orbital (Figure 5.5). Application of the symmetry operations affords the following results:

(*a*) leave-it-alone – remains unchanged and therefore has character $+1$,

(*b*) $R_{180°}$ – changes the positive lobe for the negative one thus turning the orbital into its negative. The character is therefore *minus 1*.

(*c*) Ref. (I). The p_x-orbital lies in this reflection plane and so each 'half-lobe' is reflected into the other. The orbital thus remains unchanged by this operation and the character is $+1$.

(*d*) Ref. (II). The p_x-orbital has its positive lobe on one side of the plane and its negative lobe on the other. Reflection therefore interchanges the positive and the negative lobes. Thus the character is -1. The character set for p_x is gathered in Table 5.1.

Finally, the same set of operations is carried out on the p_y-orbital. This is perpendicular to the plane of the paper (Figure 5.5). The results are:

(*a*) Identity   character $+1$
(*b*) $R_{180°}$ character -1 (like p_x)
(*c*) Ref. (I)   character -1
(*d*) Ref. (II)   character $+1$

Notice that the effect on p_y of the two reflection operations is exactly opposite to that on p_x. This is because the p_x-orbital lies in the reflection plane (I), whilst the p_y-orbital lies in plane (II) which is perpendicular to (I). The character set for p_y is listed in Table 5.1.

Now the characters for a given orbital describe the behaviour of this orbital when it is subjected to the four symmetry operations. This is a simple description but it would be tedious and clumsy if the whole set of numbers had to be written when we wanted to specify how, say, the $2s$-orbital behaved. Luckily a simple system of naming has been invented which facilitates this. When the character under the rotation operation $R_{180°}$, is $+1$ the character set is said to be of symmetry type 'a', and when this character is -1, the character set is called 'b' type. These labels 'a' and 'b' are further given subscripts '1' or '2' determined by the characters of the operations Ref. (I) and Ref. (II), e.g. if we have a set of characters which are all $+1$, they form a representation which is 'a_1' type. The way in which these labels are used as a 'shorthand' description of the character sets is shown in Table 5.1.

The $2s$-, $2p_x$-, $2p_y$- and $2p_z$-orbitals of oxygen have now been specified by their behaviour under the four symmetry operations: in this the $2p_z$-orbital has exactly the same type of symmetry (a_1) as does the $2s$-orbital. The oxygen atom has now been completely dealt with and the first stage of the treatment is complete.

The next stage is to set up appropriate linear combinations of the $1s$ atomic orbitals of the hydrogen atoms and then to find their symmetries. The only possible linear combinations of the two $1s$-orbitals are precisely the same as for the hydrogen molecule itself (Chapter 4), i.e. $(1s_A + 1s_B)$ and $(1s_A - 1s_B)$. These are depicted in Figure 5.7. It is crucial to keep in mind that a linear combination is *one function*. It looks rather a peculiar function because it is based on *two* centres whereas in 'everyday' mathematics and, of course for atoms, we usually deal with functions sited on one centre only. This need not cause alarm; we do not need to know about the analytical form of, nor how to plot, such a function, all we want to study are its symmetry properties. These can be derived in just the same way as those of the oxygen $2s$-, $2p_x$-, $2p_y$- and $2p_z$-orbitals, i.e. we apply the symmetry operations and record (Table 5.2) what happens to the signs of the L.C.A.O. when this is done. Figure 5.7 will help in following this procedure. Taking Figure 5.7 (*a*) first:

Table 5.2 *Effect of operations on linear combinations of hydrogen $1s$ orbitals*

	Identity	$R_{180°}$	Ref. I	Ref. II	Symbol
$(1s_A + 1s_B)$	$+1$	$+1$	$+1$	$+1$	a_1
$(1s_A - 1s_B)$	$+1$	-1	$+1$	-1	b_1

The component $1s$-orbitals stay in their places when the identity operation is applied and so, because the orbital is left unaltered, the character for this operation is $+1$. Rotation by 180° about the main z-axis causes $1s_A$ and $1s_B$ to change places. But because they

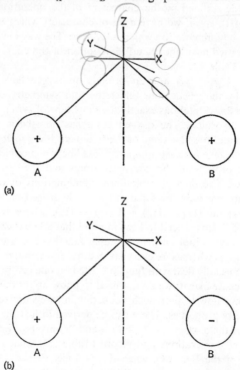

(a)

(b)

FIGURE 5.7. Linear combinations of hydrogen orbitals

both have positive signs the linear combination still looks as it did before the operation. Hence the character is again $+1$. When reflection operation (I) is applied, the hydrogen components again change places but both are positive so the character is still $+1$. So also with reflection operation Ref. II. The complete character set appears in Table 5.2 and the linear combination $(1s_A + 1s_B)$ is thus assigned to the symmetry type a_1. Working in the same way on the second of the linear combinations, Figure 5.7 (*b*), a second set of characters results, i.e.

   (i) Identity, $+1$ (as always).

  (ii) $R_{180°}$, this changes the $+$ sign for the $-$ sign and so changes the sign of the whole L.C.A.O. The character is -1.

 (iii) Reflection in Plane (I) leaves the signs unaltered and the character is $+1$.

 (iv) Reflection in Plane (II) interchanges the $+$ and $-$ signs and the character of the operation is -1.

These are listed in Table 5.2. The third stage of the analysis is now reached, i.e. construction of the molecular orbitals for the *whole* molecule. This is effected by overlapping the linear combinations of hydrogen orbitals with the orbitals on the central atom, taking cognizance of the rule that only orbitals of the same symmetry can overlap and give rise to bonds. Those which do not have the same symmetry must be orthogonal to each other and the overlap between them is zero. No bonding can be achieved by overlap between two such orbitals. When a L.C.A.O. orbital form is combined or 'overlapped' with an appropriate central atom function two new orbitals are generated. These are m.o's for the whole molecule, i.e.

$$(\text{L.C.A.O. orbital}) \pm \lambda \, (\text{central atom orbital}) \qquad (5.1)$$

The reason for having to take both $+$ and $-$ signs has been encountered in previous contexts. Moreover it has been established that the new orbital with the $+$ sign is bonding whilst the other is antibonding. The coefficient 'λ' arises because there is not necessarily a $1:1$ contribution from the two orbitals.

It is extremely important to realize the significance of nodes in polycentric molecular orbitals. A node between two atoms signifies that there is zero electron density between them (Chapter 4) and so the two atoms tend to repel each other; the term antibonding stems from this property of the orbital. Now it is quite clear in simple systems like hydrogen that there can be only one possible node in an antibonding wave function. In more complex polyatomic molecules the picture is more intricate, there are many molecular orbitals and some will be 'mainly bonding', i.e. few or no changes of sign in the L.C.A.O., and some will be chiefly of antibonding character. The number of nodes in the molecular orbital frequently aids in ordering the levels in the energy scale and is particularly useful when the electronic system under discussion involves atoms which are of one kind. There is often a close analogy to the atomic hydrogen case where both the energy and the number of nodes in the wave function depend only on n. Of course we should not expect that in systems as electronically complex as are molecules such a simple rule would be rigid. However, it will often turn out to be of great value in constructing energy-level diagrams.

The whole set of molecular orbitals for water may now be obtained, thus affording an energy correlation diagram for the molecule, somewhat like those for diatomic molecules in Chapter 4. A useful convention in building up the correlation diagram is to write the energy levels of the central atom orbitals on the *left-hand side* of a diagram and the linear combination orbitals on the *right*. To aid in clarification, the molecular levels are connected by lines to the atomic levels from which they originate. To distinguish between the bonding and antibonding levels of the same symmetry in a correlation diagram, the upper, antibonding, level is marked with a superior asterisk, e.g., a_1^*, b_1^*. Figure 5.8 illustrates the complete correlation diagram for water. The atomic energy levels of the oxygen atom are placed on the left-hand side together with their symmetry symbols. The two lines on the right-hand side represent the energies of the two linear combinations of the hydrogen $1s$-orbitals (a_1 and b_1). Now the $2s$ and $2p_z$ oxygen orbitals were shown to behave in the same way towards symmetry operations (Table 5.1): the linear combination ($1s_A + 1s_B$) also is of a_1 symmetry. The L.C.A.O. can therefore overlap with *both* the $2p_z$- and $2s$-orbitals of oxygen giving rise to m.o's of the whole molecule.

Now these *three* initial orbitals must give rise to *three* new molecular orbitals. One of these new m.o's is a bonding orbital (energy level I), the second one is an antibonding one (level VI). The third is neither strongly bonding nor strongly antibonding but is said to be a *non-bonding* orbital, i.e. it is a molecular orbital which is very close in energy to the initial level of the atomic orbitals. In the case of water it lies near the energy of the oxygen $2p$-orbital (level III in the correlation diagram). The $2p_x$-orbital (b_1) can combine with the linear combination ($1s_A - 1s_B$), which is also b_1, and again bonding and antibonding combinations are obtained. These are levels II and

V on the correlation diagram. The other $2p$-orbital on the oxygen ($2p_y$) is of type b_2 and has no symmetry match with a linear combination orbital. It cannot, therefore, bond with the hydrogen atoms at all. Hence it constitutes a second non-bonding orbital and remains

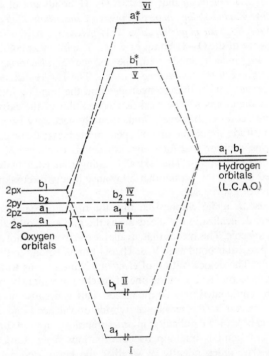

FIGURE 5.8. Correlation diagram for H_2O

at essentially the same energy level in the molecule as it was initially in the atom (level IV). The energy-level sequence is now set up and as the final act, the electrons are fed two by two into the levels. The $1s$ electrons of the oxygen were left out of the reckoning so there are only $2s^2 2p^4$, i.e. six electrons to consider. From this atom there is one electron from each of the hydrogens resulting in a total of eight. These then fill up the first four orbitals from the bottom in the energy scheme.

Several points of interest arise from this energy-level scheme. The two lowest levels containing four electrons are bonding and correspond to the two electron-pair bonds of valence-bond theory. An apparent difference is that the two m.o's are at different energies but this *does not mean* that the two O—H bonds are of different strengths. *Each bonding m.o. encompasses the whole molecule and contributes to both bonds equally.* This gives rise to the observed equivalence of the O—H bonds of water. The a_1 linear combination $(1s_A + 1s_B)$ which bonds with the oxygen $2s$ *also brings about bonding between the two hydrogen atoms.* The H—H distance is, of course, much longer than in hydrogen and the bonding interaction between the atoms will be weaker. Recognition of the existence of this interaction is, however, fundamentally necessary because our whole attitude should be one of open-mindedness; there should be no *a priori* decision that the water molecule is stable purely because of its two OH bonds. The 'extra' bonding that each hydrogen is involved in calls for yet another allowance in any discussion of the 'valency' of hydrogen.

It must be again emphasized that the m.o. approach envisages complete delocalization of electrons as the m.o's are orbitals of the *whole* molecule. This means that 'available electrons' are not assigned to any particular bond as in V.B. Theory but *all* belong to the whole molecule. The delocalization of electrons is indigenous to the m.o. approach to molecules and is one of its chief advantages in understanding conjugated organic molecules. This is because no limiting 'fixed' structures (the resonance hybrids of valence-bond theory), which can lead to difficulty in interpretation, are set up. Each equivalent bond is built up by contributions from 'static' m.o's. This is true independently of whether the bond incorporates an electron pair or not. The m.o. approach is of less value for molecules such as paraffin hydrocarbons which can, more simply and just as fruitfully, be considered as being made up essentially of an aggregate of localized electron-pair bonds. Electron delocalization plays a minor rôle and the flexibility of m.o. theory is barely required, e.g. in the molecule $H_2C{=}CH{-}CH_2{-}CH_2{-}CH{=}CH_2$ there is little evidence for delocalization of electrons from one π-bond to the other.

The heteropolarity of a bond is reflected in the coefficients (λ) of the m.o.'s. In water, the $2s$- and $2p$-orbitals on the isolated oxygen atom are lower in the energy scale than the $1s$ of the hydrogen atoms. This is mainly because the greater effective nuclear charge on oxygen attracts electrons in these orbitals more strongly and means that, in a linear combination of the form of eqn. (5.1), the oxygen and the two hydrogen orbitals do not contribute equally. Hence the two electrons in a bonding molecular orbital will not be evenly distributed between the oxygen and the two hydrogen atoms. When both bonding orbitals are taken into consideration more charge will be found to accumulate on the oxygen atom. Since each hydrogen atom has an initial charge of one electron, when bonding it must retain less than this and so the hydrogen atoms in H_2O are to some extent positively charged. Complementarily the oxygen atom accumulates excess negative charge and the overall polarity of the molecule is,

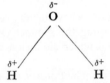

The 'δ' symbols represent small charge separations. This explains why water molecules in liquid or in ice are so strongly *hydrogen bonded* to each other: the δ^- on oxygen is attracted to the δ^+ on the hydrogens of neighbouring molecules and a macrostructure results.

The two non-bonding pairs of electrons of H_2O are essentially localized on the oxygen atom. These are the electrons which enable the water molecule to form coordinate bonds with electron-acceptor molecules, e.g. many inorganic salt hydrates contain water linked to the metal by coordinate bonds. Coordination can be described *formally* as electron donation from the donor molecule to the acceptor. The principles formulated here will facilitate description of the final coordination compound.

A full analysis of water has been carried out because it is an illustrative model. Since all our subsequent reasoning will be based on the principles of this one case, it is worth while now to summarize the mode of approach to this molecule. This illustrates the setting up of m.o. descriptions of the bonding in molecules with central atoms.

The steps are:

(1) Determine what the symmetry elements of the molecule are. This will usually be quite simple, indeed the ones above will usually turn out to be quite adequate. The guiding rule here is that one can always use fewer than there really are in the system. This will become clearer when in use.

(2) Pick out the orbitals of the central atom that are to be considered in the bonding, i.e. the available or 'valence' orbitals. In deciding this, ignore whether an orbital contains any or no electrons. This is because the m.o. approach does not take account of electrons until the final stages. It will usually be quite easy to decide on the valence orbitals, e.g. $2s$ and $2p$ are appropriate for first-row elements of the Periodic Table; $3s$, $3p$, and possibly $3d$ (although these are formally unoccupied) for the second-row elements, sodium to argon.

(3) Generate the set of 'behaviour numbers' or characters of the central atom orbitals using the symmetry operations. A physical model of the compound under discussion will aid considerably at this stage.

(4) Construct linear combinations of the orbitals on the external atoms. This will usually be fairly easy. The principle '(orbital or L.C.A.O. (1)) $\pm$ (orbital or L.C.A.O. (2))' (like $1s_A + 1s_B$ in H_2O) is an excellent guide. Derive the character sets defining the symmetry behaviour of these linear combinations.

(5) Combine the L.C.A.O's with the orbitals on the central atom which are of matching symmetry type. Remember that *two* orbitals (one from the central atom and an L.C.A.O. from the external groups) are being combined, and so must yield *two* molecular orbitals in each case, i.e. a bonding–antibonding pair.

(6) Build up the correlation diagram by writing the energy levels of the central atom on the left-hand side and the energies of the linear combinations on the right. The bonding and antibonding levels will then normally be below and above these initial levels respectively.

By extending the principles discussed here a little it is possible to construct bonding schemes for *any* molecular situation or trial model of a species. However, the correct technique for systems of high symmetry is rather harder to acquire and so we will not continue to develop symmetry theory further. Instead, using only the elementary principles and symmetry arguments discussed (plus a little 'fudging') we will examine the bonding in a variety of interesting molecules, some of which, like NO, do not conform readily to the simple electron-pair picture. However, for the moment let us pursue the case of H_2O further because bonding diagrams can be set up for *any* model at all and, e.g., a linear model could be assumed (Figure 5.9). The same symmetry elements will be equally appropriate for the linear model as for the angular one since these (among others) are included in the full range of symmetry elements of the former geometry. This is an example of the earlier statement that *fewer* than the maximum number of operations will suffice to work with. The set of behaviour numbers characterizing the $2s$-orbital on oxygen are all $+1$ just as for the angular model (indeed they are *always* $+1$

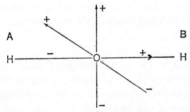

FIGURE 5.9. Linear model of H_2O – orientation of orbitals

for s-orbitals and these *always* have a_1 symmetry). The symmetry behaviour of the p_x-orbital directed towards *both* hydrogen atoms along the line of centres is readily seen to be just as it was under the angular symmetry (i.e. b_1). The two hydrogen $1s$-orbitals may, as before, be combined as $(1s_A + 1s_B)$, and $(1s_A - 1s_B)$. The former matches the $2s$ oxygen orbital in symmetry whilst the latter can overlap with p_x-orbital and so two bonding and two antibonding orbitals are produced. There are still two other p-orbitals on the oxygen atom not yet considered. These are both perpendicularly orientated with respect to the internuclear axes and, as shown earlier in Chapter 3 (Figure 3.5 (c)), they both must be orthogonal to the s-orbitals on the adjacent hydrogen atoms. This pair of p-orbitals thus does not participate in the bonding at all and contains non-bonding electrons.

There are two main differences between this situation and the angular one:

(i) The repulsions between the bonding and the non-bonding electrons are less in the angular than in the linear formulation. This is because the average distance between them is greater.

(ii) The bonding interactions between the hydrogen and the oxygen atoms are greater when the system is angular. This stems from the participation of the p_z-orbital in the bonding of the angular model, whereas it is orthogonal and completely non-bonding in the linear model. It must be remembered, however, that the *internuclear* repulsions are greater in an angular than in a linear model with the same bond length.

The whole analysis stresses the care with which one must interpret apparently simple physical factors like electron-pair repulsions. The contribution of each factor to the whole must be assessed before a decision can be reached about why a molecule is or is not stable.

Nitrogen dioxide

The molecule NO_2 is one of the 'awkward' ones which, because it has an odd number of electrons, gives trouble in the valence-bond treatment and must be represented by an assemblage of resonance hybrids. The analysis can be carried out exactly as for water. There are more orbitals involved in NO_2 but these cause no trouble if we do not lose our heads. The $2s$- and the three $2p$-orbitals on the nitrogen atom are available just as for oxygen in H_2O. The two oxygen atoms each have available their $2s$- and $2p$-orbitals and all of these could be included in the discussion. For simplicity, however, we will restrict those considered to two of the $2p$ only on each atom, i.e. one directed towards the nitrogen atom along the line of centres, and the second at right-angles to the molecular plane, i.e. sticking out of it. These orbitals are shown in Figure 5.10. The $2s$-orbitals have been neglected with their electrons and so also has the third p-orbital on each oxygen with its pair of electrons. Thus only two electrons per oxygen atom will be considered to be involved in bonding. It should be realized at this juncture that the method does not *require* the neglect of orbitals in this way, but the analysis is thereby simplified. We can return to the point later and infer how the results we get would have differed had we included them in the first place.

Now the available orbitals on the central nitrogen atom are the $2s$- and the three $2p$-orbitals just as for oxygen in water. We ignore the fact that oxygen has one more electron than nitrogen, because in this approach we deal primarily with orbitals. Because NO_2 has the same angular shape as water the orbitals on nitrogen have the same symmetry behaviour as have those of oxygen in water (see Table 5.1). The orbitals on the external oxygen atoms are different

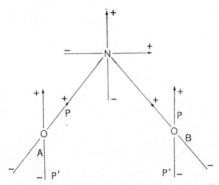

FIGURE 5.10. Orientation of orbitals in nitrogen dioxide

from the previous case in which there were only hydrogen $1s$-orbitals but there is no change in principle in the arguments applied to the present situation. Referring to Figure 5.10 (a), it is simplest to consider the pair labelled p_A, p_B first. These can be compounded into two linear combinations $(p_A + p_B)$, $(p_A - p_B)$. These linear-combination orbitals both lie in the molecular plane and on applying the operation appropriate to the angular figure, it may readily be checked that the molecular orbital $(p_A + p_B)$ is of symmetry a_1; similarly the orbital $(p_A - p_B)$ may be shown to belong to symmetry type b_1. This is a result which, in spite of using oxygen p-orbitals, duplicates the answer for the $1s$ hydrogen orbital combinations in water. We consider next the two orbitals (p'_A, p'_B) which again can be combined as $(p'_A \pm p'_B)$. Let us study these carefully as their symmetry properties are rather tricky to derive. First take the symmetric combination $(p_A' + p_B')$. The character for the identity operation is always $+1$: that corresponding to the $180°$ rotation is obtained from Figure 5.11 (a) which illustrates the effect of rotation by $180°$ on the linear combination. In studying Figure 5.11 bear in mind that p_A' and p_B' are $2p$-orbitals and so have a positive lobe sticking up out of the paper towards us and a negative lobe directed the opposite way. A rotation by $180°$ will reverse these lobes so that the negative signs are uppermost and the orbital is thus turned into its negative. The character of the rotation operation is therefore -1. Reflection in the

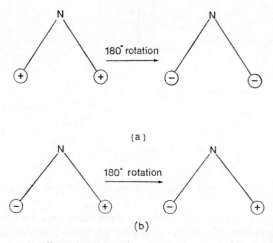

(a)

(b)

FIGURE 5.11. Effect of $180°$ rotation on p_π-orbitals of nitrogen dioxide

molecular plane will also interchange the $+$ and the $-$ signs in each case and the corresponding character is again -1. Finally, reflection across the plane perpendicular to the molecular plane (Ref. II) leaves the form of the orbital unchanged – character $+1$. The whole set of characters $(1, -1, -1, +1)$ corresponds to the set labelled b_2 (Table 5.1).

For the orbital $(p_A' - p_B')$; again the character for the identity is $+1$. Rotation by $180°$ around the central z-axis interchanges the two orbitals *but* the signs are retained (Figure 5.11 (b)). This is because the negative lobe of p_A' is rotated round and 'comes to the surface' when interchanged with p'_B. The linear combination is thus left unchanged overall by $R_{180°}$ and the character for this operation is therefore $+1$. Reflection in the molecular plane interchanges the positive for the negative lobe of each individual p-orbital and so has character -1. Reflection of the orbital across the plane (II) (Ref. II) has character -1 because of the form of the orbital ($p'_A - p'_B$). This linear combination is therefore characterized by the numbers $+1$, $+1$, -1, -1.

This is a character set we have not met before. Because it has character $+1$ for the rotation operation we label it as 'a' type and because of its behaviour towards the reflection planes it becomes 'a_2' (Table 5.1). All the orbitals have now been labelled by their symmetry behaviour but, before feeding them into a correlation diagram, it will help to recapitulate on their symmetry properties.

There is thus:

$$
\text{Nitrogen atom} \begin{cases} 2s & a_1 \\ 2p_z & a_1 \\ 2p_x & b_1 \\ 2p_y & b_2 \end{cases}
$$

$$
\begin{array}{l} \text{Oxygen} \\ \text{orbitals} \\ \text{(L.C.A.O's)} \end{array} \begin{cases} p_A + p_B & a_1 \\ p_A - p_B & b_1 \\ p'_A + p'_B & b_2 \\ p'_A - p'_B & a_2 \end{cases}
$$

Figure 5.12 illustrates how these are put together to describe the molecule. Firstly we have to take *both* the s- and p_z-orbitals on nitrogen (both a_1) to combine with the a_1 linear combination of p-orbitals. This produces three new orbitals which are

(i) the orbital of lowest energy, which, since it has no nodes, bonds all the atoms (including the oxygen–oxygen pair) together.

(ii) the complementary orbital to this which is a high-energy *antibonding* orbital, and

(iii) an essentially non-bonding one, just as in the water molecule.

Next the $2p_x$-orbital on nitrogen (symmetry b_1) bonds with the

K

b_1 linear combination of p-orbitals affording a bonding–antibonding pair which also appears in the scheme for water.

Some new features of interest now appear. In water the $2p_y$-

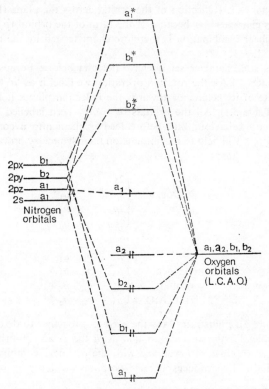

FIGURE 5.12. Correlation diagram for nitrogen dioxide

orbital on oxygen (b_2) remained non-bonding because there was no matching linear combination of the hydrogen orbitals. In NO_2 there is an L.C.A.O. of just this symmetry on the oxygens, i.e. the orbital ($p_A' + p_B'$). A new bond, not present in water, thus manifests itself. This does not lie along the line of centres because the orbitals which form it are directed perpendicular to the molecular plane. This new bond is thus of π type: all three atoms partake in this bond and its electrons are delocalized over *all three* centres. This is not a 'full' π-bond (i.e. an electron-pair bond) between any *two* atoms but a 'partial' one between them all that is both O—N *and* O—O. Thus there is a degree of double bonding throughout the whole NO_2 molecule. The longer distance between O—O compared with O—N means that the bonding between this pair of atoms is much weaker – but it is a factor which cannot simply be ignored.

Finally there is a second interesting molecular orbital formed from p_A' and p_B'. This has the form $(p_A' - p_B')$ (a_2 symmetry) and has no match on the nitrogen atom. It cannot therefore aid in holding the oxygen atoms to the nitrogen. In fact the orbital has a node between the two oxygen atoms and it does not help to bond these two together either. This orbital is therefore non-bonding with respect to the nitrogen atom and antibonding with respect to the initial oxygen levels. Five electrons originate from the nitrogen atom and two electrons from each oxygen (since we neglected the $2s$ and one $2p$ orbital of oxygen plus four electrons from each). If we feed the nine electrons into the empty levels from the bottom the final three electrons are virtually non-bonding. It is interesting that the 'last' electron inhabits an a_1 type orbital chiefly of p_σ character and centred mainly on nitrogen. This odd electron causes no trouble in assignment or interpretation because in m.o. theory levels are set up simply in order of energy and the available electrons are fed into them. This is exactly how we arrive at the electronic structure of an atom.

The molecule is thus bonded together by means of two equivalent σ-bonds built from contributions from the a_1 and b_1 σ-bonding molecular orbitals and a single π-bond delocalized over all three centres. To complete the electronic picture there are three, virtually non-bonding, electrons. The electronic structure of the nitronium ion, NO_2^+, which is the principal attacking species in organic nitrations, is easily derived from Figure 5.12 and a problem related to this ion is included at the end of this chapter.

If we had considered the $2s$- and the third $2p$-orbital on the oxygen atoms in our treatment, this would have produced a more compli-cated correlation diagram, bringing in four more pairs of electrons and four more orbitals. The overall bonding description of the molecule, however, remains essentially the same.

Molecules containing π Electrons

We are now at a stage to use the experience gained in simple symmetry applications in order to study the electronic structural features of larger molecules where a knowledge of electronic effects gives particular insight into their reactions. One of the most interesting groups of organic molecules are the unsaturated hydrocarbons and their derivatives. We will first study the simplest, ethylene, and work from this case on to larger molecules.

Ethylene

This molecule has already been described by the valence-bond theory (Chapter 3). Before applying the m.o. approach it is necessary to do a little background work. Planar molecules like ethylene, butadiene and benzene possess a common symmetry element; the molecular plane is a reflection plane of symmetry. This means that the m.o's describing the molecules must be divided into two types: (*i*) those which remain unchanged when reflected through the molecular plane and (*ii*) those which change sign when this is done. This was encountered in the NO_2 molecule where the orbitals of σ type (a_1, b_1) did not change sign on reflection in the molecular plane whereas those of π type (the b_2 and a_2 pair) did. For planar molecules this effect always obtains and it separates the σ- and π-orbitals – it therefore gives rise to the σ–π *separability condition*. The interesting physical and chemical properties of unsaturated hydrocarbons and many of their derivatives stem mainly from electrons in the π-type orbitals. The σ levels act only as a kind of 'background noise' because in most cases they lie well below the π levels in the energy scale (this emerges from the correlation diagram of NO_2 where the b_2 and a_2 bonding and non-bonding levels are higher than the a_1 and b_1 σ levels).

It is reasonable then to regard ethylene as composed primarily of a σ *skeleton* of sp^2 hybrids overlapping with the $1s$ hydrogen orbitals, as discussed in Chapter 3. The π-orbitals of the molecule are then constructed by the L.C.A.O. method from the unhybridized $2p$-orbitals perpendicular to the molecular plane. Since this can be done for all planar unsaturated hydrocarbons, we call them π electron molecules. A striking similarity to a simple m.o. case, i.e. H_2,

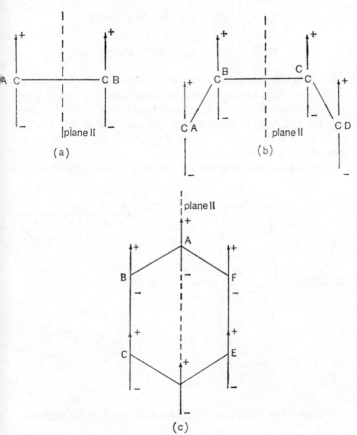

FIGURE 5.13. The σ skeleton of (a) ethylene, (b) butadiene, (c) benzene

is immediately obvious. The latter was described by L.C.A.O's built from the $1s$-orbitals centred on the hydrogen atoms. The relationship of the ethylene π system to hydrogen should be evident and so to describe ethylene, L.C.A.O.'s of the $2p$-orbitals centred on the 2 carbon atoms may be used (Figure 5.13 (*a*)). Hence we take as usual,

$$\Psi_1 = \frac{1}{\sqrt{2}}(p_A + p_B) \quad \text{bonding orbital}$$

$$\Psi_2 = \frac{1}{\sqrt{2}}(p_A - p_B) \quad \text{antibonding orbital}$$

$1/\sqrt{2}$ is the normalizing factor for the orbital and fulfils the conditions which were discussed for linear combinations in Chapter 3 (i.e. in $(C_1 \phi_1 + C_2 \phi_2)$, we must have $C_1{}^2 + C_2{}^2 = 1$ in order that the L.C.A.O. be normalized). The same factor '*C*', $1/\sqrt{2}$, is given to each $2p$-orbital because they are of the same type and centred on equivalent atoms.

Now we have generated two orbitals for the π electrons. That with

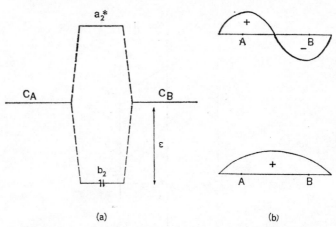

(a)   (b)

FIGURE 5.14. Correlation diagram and π-waveforms for ethylene

the positive sign is symmetric to the reflection plane II (Ref. II) (Figure 5.13 (*a*)), and has the lower energy of the two. The correlation diagram (Figure 5.14 (*a*)) sums up the disposition of the two levels about the energy zero. The latter is the energy which a $2p$-electron of carbon in ethylene would have if it were not participating in π-bonding, i.e. if it were simply isolated on its C atom. It corresponds in the hydrogen problem to the energy of an electron on an isolated H atom. The bonding and antibonding π levels are thus distributed *symmetrically around the energy zero*. The symmetry behaviour of these two orbitals is easily established by recourse to the method carried out on NO_2. The molecular orbital $(p_A + p_B)$ is unchanged by the identity and by reflection across the plane II (Ref. II). Rotation by $180°$ around z (an axis in the plane of the paper) and reflection in the molecular plane *reverses* the sign of the m.o.'s. This orbital is thus classified as b_2. By similar reasoning, the orbital $(p_A - p_B)$ may be shown to have symmetry a_2. The forms of the wave functions are shown in Figure 5.14 (*b*). The higher-energy, antibonding orbital has one node.

Butadiene

The next system of interest is *cis*-butadiene (Figure 5.13 (*b*)). In this we again assume a σ skeleton constructed on essentially the same lines as ethylene. Although no atom lies at a central point, nevertheless the molecule has precisely the same four symmetry elements as do the letters A and E and the molecules H_2O and NO_2. Now the molecular orbitals of H_2, H_2O and the π m.o.'s for C_2H_4 were made up by taking '(orbital (A) $\pm$ orbital (B))'. These are the symmetric and antisymmetric combinations yielding a bonding–antibonding pair of molecular orbitals. With slight modifications the same can be effected for butadiene; we take the $2p$-orbitals on the carbon atoms, thus

$$(p_A + p_B) \pm (p_C + p_D)$$

i.e. (one side of the molecule) $\pm$ (the other side). This affords *two* new molecular orbitals. There should of course be four because there are four 'starting' atomic $2p$-orbitals. The two others are not far to seek. They are;

$$(p_A - p_B) \pm (p_C - p_D).$$

These are the four molecular orbitals describing the π-electron system of butadiene.* Electrons are thus delocalized over the whole of the molecule conferring double bond character on each C—C bond. The delocalization phenomenon is thus described quite

* Orbitals such as $(P_A - P_B) + (P_C + P_D)$ are inappropriate – see Problems.

naturally using m.o. principles without recourse to writing resonance hybrids to account for the partial double-bond character of the central C—C bond in this compound. The energy-level scheme is illustrated in Figure 5.15 (a). There are four levels two of which are bonding (energy below that of an electron in an isolated p-orbital)

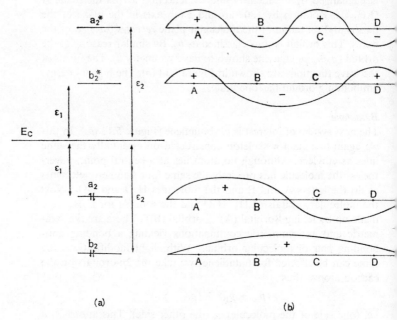

(a)   (b)

FIGURE 5.15. Correlation diagram and π-waveforms for butadiene

and two antibonding. The four π electrons, one from each carbon atom, fill the former completely. The two antibonding levels are the same distance above the energy zero as the bonding ones are below it (the energies ε_1 and ε_2 are equal).

The forms of the orbitals are shown in Figure 5.15 (b). The arrangement of the orbitals in the energy scale follows from the number of nodes in the orbitals; the molecular orbital with all positive signs has no nodes and lies at the lowest energy, whereas the orbital with alternate positive and negative signs has the highest energy. The latter orbital exhibits a sign change between each pair of atoms and so is antibonding between all of them.

To classify the orbitals we use the four defined operations. It is left to the reader to show that they have the symmetry labels in Figure 5.15 (hint: draw four models of the molecule and put appropriate signs on the atoms for each m.o.). In conclusion it should be re-emphasized that these four m.o's encompass the whole molecule and the two bonding pairs of electrons contribute both to each 'full' π-bond and to the 'partial' π-bond.

Benzene

Finally, employing the same principles, we shall tackle benzene. This is the conjugated system *par excellence* and, as is well known, its shape is that of a regular hexagon. Although benzene has really many more symmetry elements than the four used hitherto, nevertheless it does possess these and we can work quite satisfactorily with these only. An axis through one of the atoms is taken as the 180° rotation axis (Figure 5.13 (c)) and it is assumed as before that the σ electronic system is adequately described by a set of sp^2 hybrids on each carbon atom linked both to the others in the ring and to the six hydrogen atoms. No further reference to the σ system need be made. It is the set of six $2p$-orbitals sticking out of the molecular plane and their six associated electrons which confer on benzene most of its interesting chemical and physical properties.

Two 'central atoms' may be designated in the π system; one at the 'top' and one at the 'bottom' of the molecule (A and D). Their single $2p$-orbitals have identical symmetry behaviour which can easily be shown to be of b_2 type. Linear combinations of the remaining four orbitals (p_B, p_C, p_E, p_F) on either side of the central atoms are taken (cf. butadiene).

These are:

$$(1) \quad (p_B + p_C) + (p_E + p_F) \qquad b_2$$
$$(2) \quad (p_B + p_C) - (p_E + p_F) \qquad a_2$$

and

$$(3) \quad (p_B - p_C) + (p_F - p_E) \qquad b_2$$
$$(4) \quad (p_B - p_C) - (p_F - p_E) \qquad a_2$$

The symmetries of these orbitals are the same as those of butadiene. We can now proceed further and combine them, where appropriate, with the two 'lone' orbitals A and D. Now the two a_2 linear combinations (L.C.A.O's) (2) and (4) cannot combine with A and D because of their different symmetry behaviour. These two L.C.A.O's are, therefore, two final molecular orbitals of benzene and are shown

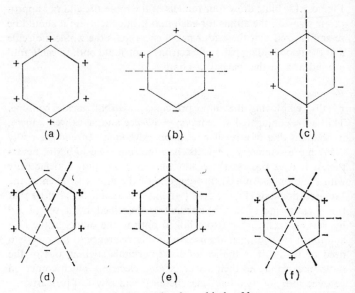

(a)   (b)   (c)

(d)   (e)   (f)

FIGURE 5.16. π molecular orbitals of benzene

in Figure 5.16 (c) & (e). The other two orbitals of b_2 symmetry can combine with p_A and p_D and so give rise to

$$(p_B + p_C) + (p_F + p_E) \pm (p_A + p_D)$$

and

$$(p_B - p_C) + (p_F - p_E) \pm (p_A - p_D).$$

The forms of these orbitals are also illustrated in Figure 5.16; the dotted lines represent the position of the nodes in the wave functions. All six of the benzene π-orbitals formed from the six initial atomic orbitals have now been obtained. The sequence of the number of nodes allows their placement in an energy order as follows: Orbital (a) has no nodes between atoms and hence lies lowest in energy. Orbitals (b) and (c) both have one node and these two levels have the *same* energy, i.e. they form a doubly degenerate pair. Orbitals (*d*) and (*e*), each having two nodes, are also degenerate and fall next in the energy sequence. Finally, in orbital (*f*) there is a node between each pair of atoms and hence this orbital is highest of the six in the energy scale. The energies of the six levels are shown in Figure 5.17. The energy levels are equally spaced around the energy zero, just as in ethylene and butadiene and the six electrons of the

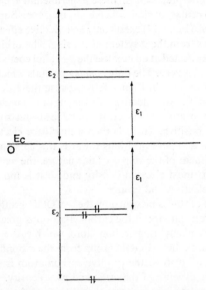

FIGURE 5.17. π energy levels of benzene

system fill up all the first three bonding levels. The electrons described by these bonding orbitals are delocalized all round the ring which results in all atomic positions and bonds in benzene being equivalent. The known properties of the compound are consistent with such a picture. It is important to understand the electronic make-up of benzene fully since it occupies an extremely important position in organic chemistry. Substituent effects in aromatic chemistry are understood properly only on a basis of a detailed knowledge of the benzene structure.

Substituent Effects in the Benzene Ring

It has been well established experimentally that certain substituents in benzene affect the position of further substitution in the ring, e.g. monosubstitution by chlorine, the hydroxyl- or the amino-group directs further substitution by negative-centre-seeking groups into the *o* and *p* positions, whereas substitution of the nitro-group in a benzene ring causes further substitution to occur in the *meta* position. An interpretation of these experimental facts is afforded by the π molecular orbital scheme of benzene. Suppose that a substituent at A (Fig. 5.13) exerts only an inductive effect, i.e. it withdraws electrons from the σ system of the ring. The orbital (*c*) in Fig. 5.16 cannot be affected at all because the *p*-orbital coefficient on atom A in this orbital is zero. The only *bonding* orbitals which can be perturbed are (*a*) and (*b*) in Figure 5.16 because these do have coefficients on atom A. An inductive substituent draws σ electrons away from the carbon atom to which it is attached and so causes it to become more positive. There is then a flow of π electrons, mostly from the *ortho* and *para* positions towards this carbon. Thus when the first substituent of the ring is of this nature, the *meta* position in the ring *retains* most electron density and so it is most susceptible to attack by electrophilic groups.

An example of this is provided by the —COOH group. In benzoic acid the electronegativity of the carboxyl-group is greater than that of the adjacent phenyl ring carbon atom, which causes electrons to be drawn towards the —COOH group from the σ-bond between the two. An electron drift to the positive carbon atom is set up in the π system at the expense of the *ortho* and *para* density. Further substitution by electrophilic reagents then occurs at the *meta* position, e.g. nitration yields *m*-nitrobenzoic acid.

Certain other substituents possess a *p*-orbital which is orientated at right-angles to the phenyl ring and contains a lone-pair of electrons. Examples of this type of substituent are —Cl, —OH and —NH$_2$ (Figure 5.18). The *p*-orbitals of these substituents can enter into bonding with the ring π system *via* the ring orbitals of b_2 symmetry, i.e. those which have coefficients for the adjacent phenyl ring carbon atom. The effect of this interaction is that electrons are delocalized from the substituent into the π-orbitals of the ring and appear at the *ortho* and *para* positions. These positions are thus rendered susceptible to attack by electrophilic reagents, e.g. the nitro-group, which attacks as NO_2^+, readily substitutes into phenol producing first *ortho-* and *para*-nitrophenol and finally 2,4,6-

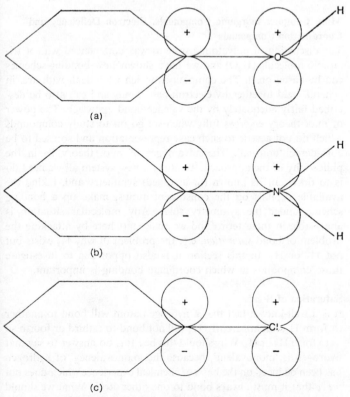

FIGURE 5.18. p_π interaction in phenol, aniline and chlorobenzene

trinitrophenol. In organic chemistry, this effect is usually called the *mesomeric effect*.

The principles acquired through the study of both water and benzene may be taken in combination to describe many other organic molecules. Diphenyl ether is one such. The π-orbitals of this molecule are constructed by taking the linear combinations of benzene previously discussed and forming new m.o's from these on the principle of ring (1) $\pm$ ring (2). The appropriate members of this set then overlap with the $2p_y$ π-type orbital on the oxygen atom.

More Complex Inorganic Compounds. Electron Deficiency and Coordination Compounds

The discussion of polyatomic systems was commenced with a triatomic molecule (H_2O) and we have shown how bonding schemes can be based on it. The compounds which were dealt with are, in general, held together by electron-pair bonds and can also be described fairly adequately by the valence-bond approach. The power of m.o. theory emerges fully when we go on to study compounds which do not accede to such easy representation and so used to be considered 'unusual'. The advantages of m.o. theory lie in the philosophy of the approach. To study a new system all we need do is to draw out its known (or assumed) symmetry and, taking the available orbitals on the constituent atoms, make up a bonding scheme under the symmetry rules. Any molecular situation is analysable in these terms and we shall start here by attacking the problem of bond *saturation*, e.g. the problem of why H_2 exists but not H_3 or H_4. In this section it is also opportune to investigate those compounds in which coordinate bonding is important.

Saturation of Bonds

It is a well-known fact that a hydrogen atom will bond to another to form H_2, but apparently H_2 will not bond to a third or fourth so as to form H_3 or H_4. Why should this be? It is no answer to say that hydrogen is 'monovalent', because the 'monovalency' of hydrogen has been deduced on the basis of chemical experience which does not *prove* that it must always bond to one other atom. What we should do is to regard the possibility of molecules such as H_3 or H_4 with an open mind and to try and *discover* possible reasons for their non-existence. In the light of our discussion of the covalent bond there is, after all, no *a priori* reason for the non-existence of such entities. The final situation simply depends on whether the overall attractive bonding forces outweigh the overall repulsive forces and whether the valency electrons are bonding, non-bonding or antibonding.

The bonding characteristics of the molecule H_3 will first be examined. This will also serve as a basis for a later interesting

(a)

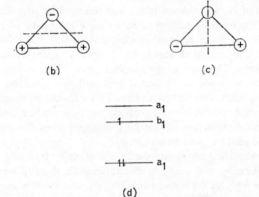

(b)   (c)

(d)

FIGURE 5.19. Orbitals and correlation diagram for H_3

modification. We assume an equilateral triangular geometry for the molecule (Figure 5.19) and so the bonding problem is closely allied to that of water. The hydrogen $1s$-orbital A has a_1 symmetry and this can overlap and bond with the symmetric linear combination orbital (B + C) yielding (B + C) + A (Figure 5.19 (a)). The complementary combination $(B + C) - A$ (Figure 5.19 (b)) is the corresponding antibonding orbital; it also has a_1 symmetry. The third, and only other, linear combination $(B - C)$ (b_1) cannot bond to the $1s$-orbital A. This is because the latter does not change sign either on $180°$ rotation or reflection across plane II whereas $(B - C)$ does. The nodes in the three wave functions lead to the energy-level diagram for the system shown in (Figure 5.19 (d)). There are three electrons, one from each hydrogen and two occupy the lowest bonding orbital. The third goes into the next orbital $(b_1$ symmetry) which is non-bonding. Hence the molecule H_3 would be held together by two bonding electrons with the third not adding to the stability at all. On average this is less than one electron per bond. The bonding forces are no more potent than in H_2 itself and, moreover, to 'pay' for this new geometry two additional H—H internuclear repulsions together with electron–electron repulsions must be suffered. Overall it 'scores' over H_2 only in its possession of an extra non-bonding electron. On the grounds of previous arguments, it is obvious why the molecule would be unstable. The energy-level diagram of H_3 also demonstrates that the stability situation would not be improved by adding or removing electrons to give H_3^+ or H_3^-. The usage of correlation diagrams in this way often yields invaluable information about, and insight into, the behaviour of ionic species. The study of the species H_3 has shown why the normal hydrogen molecule is 'saturated' and will not bond on another hydrogen atom. This will be confirmed later for the H_4 case. Valency saturation, however, is not a property which should be *assumed* for a particular atom, but should be discovered from first principles. This will lead to its understanding.

It is interesting to modify the H_3 situation by inserting a fourth atom centrally and studying the differences thereby introduced. Suppose we take the above situation, H_3, and insert a *boron* atom into the centre of the triangle. We will then have borane, BH_3, although the molecule is not usually drawn with bonds joining the hydrogen atoms. Now there is no *a priori* reason to draw lines between the hydrogen and the boron atoms. It might be found that the most important stabilizing forces in BH_3 were between the hydrogen atoms and not between hydrogen and boron, except, of course, that we have already shown that the H—H bonding contribution to stability is not likely to be very important. Nevertheless it is a good plan to remain open-minded about 'bonds' if we are to appreciate the elegant arguments regarding more complex molecules. We shall now consider BH_3 in the context of the related boron hydrides.

Boron Hydrides and Methane

The nature of the bonding in the boron hydrides has been the subject of much puzzled speculation since they were first prepared by Stock in 1912. The basic difficulty is that (except for BH_3) there are not sufficient electrons for there to be a two-electron bond between *each* pair of bonded atoms, e.g. the simplest neutral compound, B_2H_6, requires fourteen valence electrons if it is to have an ethane-like structure. Although diborane is now known not to possess such a structure, we still cannot hope to understand it if we try to explain the bonding in terms of electron pairs. On the basis of an m.o. approach we can create a satisfactory picture of the electronic structure of the compound. This affords a good basis for all the others. Before tackling B_2H_6, however, it will be a good idea to proceed with borane which was mentioned above. A quantity of BH_3 is energetically unstable towards dimerization (to B_2H_6) although a single molecule is not thought to be *intrinsically* unstable. In order to attempt to discover the reason for its instability towards dimerization we must study both BH_3 and B_2H_6.

Borane has planar, trigonal symmetry (Figure 5.20). There is a $180°$ rotation axis along each of the lines joining boron and hydrogen and, in this way, the system closely resembles water in symmetry. Of course, like benzene, it really has more symmetry elements than does water (see back to beginning of chapter, Figure 5.1), but a perfectly adequate m.o. description for our purpose may be set up under the kind of scheme appropriate to H_2O. The $2s$- and the three $2p$-orbitals are available on the boron atom (the $1s$-orbital is, as usual, left out of the reckoning) and these orbitals are readily labelled using the four familiar operations. The labels of the orbitals are the same as they were for O in H_2O or N in NO_2, i.e.

D	$2s$	a_1
D''	$2p_x$	b_1
D'	$2p_z$	a_1
	$2p_y$	b_2

L

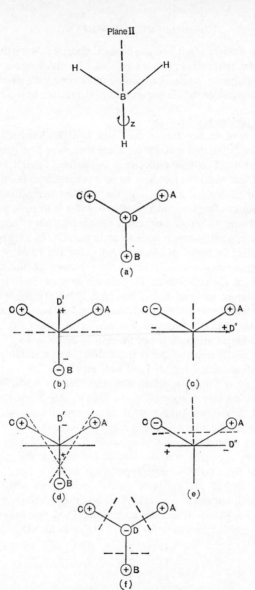

FIGURE 5.20. Symmetry and molecular orbitals of BH₃

The orbitals A, B and C are just H $1s$-orbitals. 'B' differs from the others in that it lies on the z-axis and constitutes a second 'central atom'. It is therefore considered together with the boron orbitals and adds a single $1s$-orbital of symmetry a_1. The method of proceeding is as before; two L.C.A.O's can be set up from orbitals $1s_A$ and $1s_C$ which are

$$(s_A + s_C) \qquad a_1 \text{ symmetry}$$
$$(s_A - s_C) \qquad b_1 \text{ symmetry}$$

These are now combined with the orbitals of the central boron atom bearing in mind that 'orbital (1) $\pm$ orbital (2)' bonding and antibonding combinations are always formed. This yields the three pairs of orbitals (Figure 5.20 $(a-f)$), see opposite page.

(Figure 5.20)

(a)	$(S_A + S_C) + (S_B + S_D)$	bonding, no nodes
(f)	$(S_A + S_C) + (S_B - S_D)$	antibonding, three nodes
(b)	$(S_A + S_C) - (S_B - D')$	bonding, one node
(d)	$(S_A + S_C) - (S_B + D')$	antibonding, two nodes
(c)	$(S_A - S_C) + D''$	bonding, one node
(e)	$(S_A - S_C) - D''$	antibonding, two nodes

The $2p_y$-orbital of boron sticks out of the plane of the molecule and so for symmetry reasons cannot form bonds with either of the L.C.A.O's of $1s_A$ and $1s_C$. Hence it is constrained to remain nonbonding. Notice that the two m.o's (c) and (e) contain no contribution from the hydrogen $1s$-orbital B. This is because the p_x-orbital of boron (D'') changes sign by reflection across plane (II) but the hydrogen orbital $1s_B$ cannot. Its participation in these m.o's is thus effectively precluded. The symmetry in electron distribution in the compound is maintained by molecular orbitals in the second group: these incorporate larger coefficients of orbital $1s_B$ to compensate.

Figure 5.21 depicts the correlation diagram. The energy order is based on the number of nodes in the wave functions. There are six bonding electrons (three originate from boron and one from each of the hydrogens) and they fill the three lowest energy orbitals. The bonding picture correlates well with the valence-bond description, where three sp^2 hybrids give rise to three B—H electron-pair

bonds by overlapping with hydrogen $1s$-orbitals. It should be remembered that, in the m.o. scheme, the three orbitals do not each
describe one bond, but all orbitals contribute to each bond thereby
explaining their equivalence.

Orbitals (b) and (c) each show one node and are a degenerate
pair, and, for the same reason, so are orbitals (d) and (e). There is

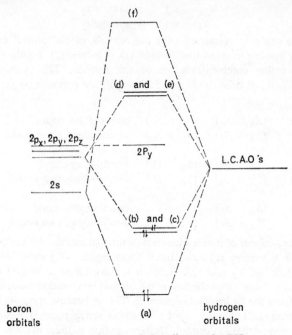

FIGURE 5.21. Correlation diagram for BH_3

a striking similarity to the benzene orbital scheme. The $2p_y$-orbital
of boron, placed centrally in the diagram, takes no part in the bonding of this molecule. The lowest bonding orbital has no nodes
whereas the highest antibonding orbital has a node between the
boron and all the hydrogen atoms.

The $2p_y$-non-bonding orbital is unoccupied and is conveniently
situated (in the energy sense) to receive any 'incoming' electrons.
It is the presence of this empty orbital which confers the very impor-

tant property of *electron deficiency* on this and many other boron compounds. This particular electronic feature, in fact, dominates virtually the whole of boron chemistry. Thus although BH_3 is *intrinsically* thermally stable it has a tendency to interact with any other molecule which will cause this empty non-bonding orbital to become bonding instead. This leads to the formation of a large number

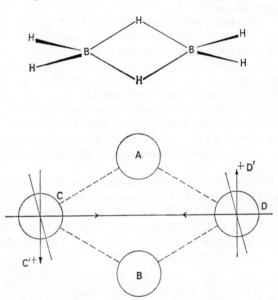

FIGURE 5.22. Structure and orbital orientation in diborane

of *coordination complexes*. One of these is the dimer of BH_3 itself, B_2H_6. This makes an interesting study because there are two few electrons in two borons and six hydrogens for eight electron-pair bonds.

Figure 5.22 illustrates the geometry of diborane and the orientation of the atomic 2s- and 2p-orbitals on the two boron atoms. As before, the axes alone represent the p-orbitals and the way in which the positive and negative axes of C' and D' are defined should be noted. The *terminal* B—H bonds are just like those in BH_3 and are not particularly important in a comparative discussion of the two. They are essentially simple electron-pair bonds and each boron contributes one electron to each of them. A boron atom has, initially, three valence electrons and so one only is available on each atom for use in the bridging system. This is the crux of the whole problem; compounds such as this with more valence orbitals than electrons are electron deficient.

The $-BH_2$ end unit is angular like H_2O and we may suppose that two of the $2p$-orbitals of each boron are fully involved in bonding with the terminal hydrogen atoms. This leaves only the $2s$- (C and D) and the third $2p$-orbital (which is always non-bonding in angular hydrides: see the energy-level diagram for water). The latter are designated as C' and D' in Figure 5.22. Hence it is a reasonable approximation to assume that these remaining four orbitals (two per boron) will be the principal contributors to the bridge bonding. By making this approximation we shall not obtain *all* the features of the bonding but the ones of greatest interest should emerge.

The orbitals to be considered are labelled A, B, C, D, C', D' in Figure 5.22. The linear combinations constructable from the boron orbitals are $C + D$, $C - D$, $C' + D'$, $C' - D'$ and they can be combined with the bridge hydrogen $1s$-orbitals A and B. (The latter could be thought of as two 'central atoms'.) Taken all together there are six resultant orbitals which are

$$(C + D) - (A + B) \qquad a_1$$
$$(C + D) - (A + B) \qquad a_1$$
$$(C - D) \qquad b_1$$
$$(C' - D') + (A - B) \qquad a_1$$
$$(C' - D') - (A - B) \qquad a_1$$
$$(C' + D') \qquad b_1$$

The orbitals are labelled according to their symmetry behaviour to rotation around the axis of the two hydrogen atoms. (It is interesting that four of these orbitals are closely similar to the π molecular orbitals of butadiene.) This similarity of apparently unrelated compounds is one of the elegant and simplifying features of molecular orbital theory; the forms of the orbitals describing a molecule are determined completely by the *symmetry* of the situation.

Figure 5.23 shows the form of the six orbitals. Now one electron was left on each boron atom for the bridging bonds, one is provided by each hydrogen atom and so there are four to feed into the molecular orbitals. The energy order of the orbitals is given in Figure

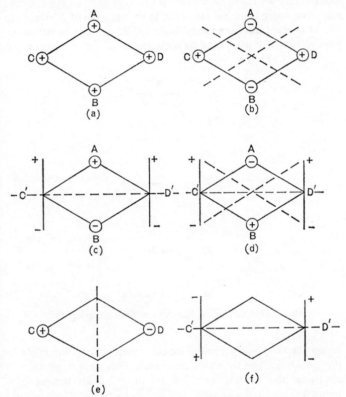

FIGURE 5.23. Orbital forms for the hydrogen bridge in diborane

5.24, the two lowest (Figure 5.23 (*a* & *c*)) are seen to be satisfyingly bonding and accommodate the four 'bridging' electrons. In diborane, therefore, *all* the orbitals of both boron atoms and the hydrogens are involved in bonding, whereas in the monomer, BH_3 (Figure 5.20), there was one empty *non-bonding* atomic orbital left on the boron. The m.o. approach even at this approximate level thus supplies a satisfactory explanation of the existence of a molecule even though there are insufficient electrons to allot two to each bond. Other boron hydrides are amenable to similar treatment although there are many more orbitals involved and consequently the analysis is more complex. However, in the example of diborane the crucial concept has been established: a hydrogen can be covalently bonded to two borons (and to some extent to the other hydrogen

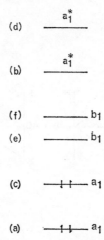

FIGURE 5.24. Correlation diagram for diborane bridging bonds

across the ring) solely because we can construct bonding m.o's for the system sufficiently stable that, even with limited numbers of electrons, the binding energies still outweigh all the repulsion terms.

The example of diborane just discussed is a special case of a *coordination compound* of borane. A wide range of such compounds is formed between other boron compounds acting as electron acceptors and electron-rich donors such as ammonia. One of the earliest of these to be discovered was $BF_3 . NH_3$ (Figure 5.25). Trigonal, planar boron compounds are generally electron acceptors by virtue of an empty atomic orbital like the one which appeared in the correlation diagram for borane. In other boron compounds this orbital is, to a greater or lesser extent, non-bonding. Coordination of a donor molecule constrains the acceptor to take up a pyramidal configuration and hence the non-bonding orbital is forced to take part in the new σ-bond to the donor. This is essentially how the coordinate bond is described.

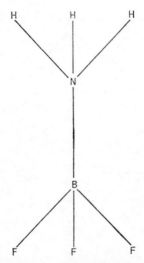

FIGURE 5.25. Boron trifluoride–ammonia complex

A full m.o. treatment of the bonding is not difficult to carry out – to do this the two component molecules are considered together. We shall not take this further because the essential point has already been revealed, i.e. the p-orbital which is non-bonding in planar compounds becomes fully involved on coordination in σ-bonding to the donor.

The next subject for study will develop an earlier theme; the H_3 molecule was unstable but was easily converted into electronically stable BH_3 by inserting a boron atom in the centre. What about the molecule H_4? Figure 5.26 shows an assumed model of the system

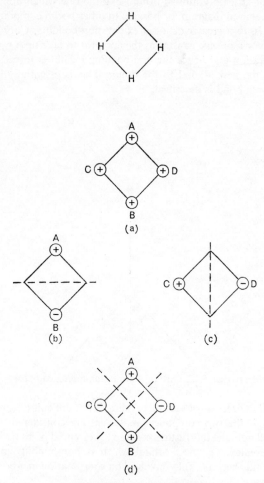

FIGURE 5.26. Orbital forms for H_4

and the forms of the four molecular orbitals which can be constructed by symmetry from the four component $1s$-orbitals. It should be noticed that these resemble closely both those of diborane and of

butadiene. To derive the correlation diagram the orbitals are arranged in order of nodes (Figure 5.27). The two orbitals which
have one node each form a degenerate pair and are slightly antibonding. There are four electrons available of which two are
bonding. The others are allotted one to each of the degenerate
orbitals. They have the same spin and the situation is thus parallel
to that encountered before in the case of O_2. We conclude that in
H_4, just as in H_2 itself, there are still only two bonding electrons
and so the same situation as found in H_3 emerges again – the molecule H_4 has no more bonding electrons than has H_2 but the electronic

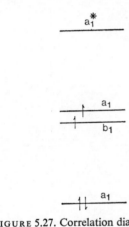

FIGURE 5.27. Correlation diagram for H_4

and internuclear repulsions are greater. It is hardly surprising that
the system is unstable – constituted as two individual H_2 molecules
there would be two bonding pairs of electrons and less nuclear
repulsion.

However, H_3 was modified successfully by placing a boron atom
at its centre so that an intrinsically stable compound was produced.
The next step is obvious: the same technique can be applied to H_4
by locating a carbon atom in the centre of the H_4 ring to produce
CH_4. The linear combinations of the four hydrogens are the same
as if the carbon atom were not present. To study planar CH_4 all
we need do is to match the carbon atomic orbitals to the H_4 ring

molecular orbitals (Figure 5.28). The $2s$-orbital of carbon matches the a_1 L.C.A.O. (Figure 5.28 (a)) whilst the $2p_x$- and $2p_z$-orbitals match the combinations 5.28 (d) and 5.28 (c) respectively. The linear

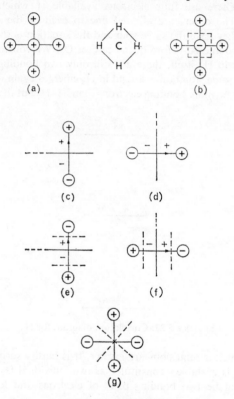

FIGURE 5.28. Orbital forms of planar methane

combination (Figure 5.28 (g)) has no match with any of the central atom orbitals and so is essentially non-bonding. A second non-bonding orbital is the $2p_y$-atomic orbital on carbon which is directed out of the molecular plane. This orbital changes sign on reflection in the molecular plane; thus it cannot bond with any combination of hydrogen $1s$-orbitals because these do not undergo such a change.

Figure 5.29 shows the energy order of the orbitals for the system. The position of the bonding levels is again determined by the nodes. If the electrons (there are eight of these, four from carbon and four from hydrogen) are allotted two by two to the energy levels then there are three bonding pairs of electrons and one non-bonding pair. This means that there are only six bonding electrons with

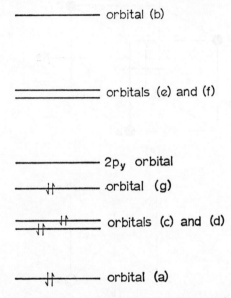

FIGURE 5.29. Correlation diagram for planar methane

which to hold all four hydrogens to the central carbon. The 'last' pair of electrons occupies a non-bonding orbital spread out over the four hydrogens. Thus, although the H_3 species could be rendered stable by inserting a boron atom in the centre, the same effect cannot be achieved by putting a carbon in the middle of the H_4 ring. Planar CH_4 possesses a pair of 'useless' electrons which do not offset the increased nuclear repulsions which ensue from the introduction of a highly charged carbon nucleus into the system. The system as it stands would be more stable as two hydrogen molecules and a carbon atom, or, of course, in some other shape.

This last point provides the clue: a planar form is not the only possible shape for CH_4. The four hydrogen atoms could alternatively be arranged at the corners of a regular tetrahedron, with

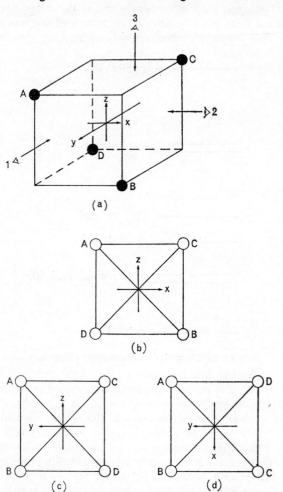

FIGURE 5.30. Geometry and projections of a tetrahedron

carbon lying at the centre. This is the experimentally known geometry of methane and it is worth re-examining the molecule theoretically to see if symmetry arguments help to show why it does take up this shape. The symmetry of the tetrahedral situation is higher than previous cases and since it is non-planar, it is more difficult to appreciate the forms of the orbitals properly. However, the effort is well worth while and Figure 5.30 illustrates the tetrahedron and the way in which the geometry is related to that of the cube. The hydrogen atoms A, B, C and D are arranged at four of the corners of the cube and as before contribute $1s$-orbitals to bonding. The forms of the orbitals are more readily understood if we look at the tetrahedron from the three vantage points 1, 2 and 3. From (1), for example, the x- and z-axes on the central atom will be seen face on (remember these axes are equivalent to the p_x- and p_z-orbitals).

The carbon $2s$-orbital can bond with a linear combination of hydrogen orbitals with all signs positive (Figure 5.31 (a)) and corresponding to this there is, of course, an antibonding orbital (Figure 5.31 (b)). The p-orbitals of carbon can be separately matched up to linear combinations of the hydrogen orbitals. Figure 5.31 (c) depicts the L.C.A.O. which matches the $2p_z$-carbon orbital (in this diagram, the x-axis is drawn in to identify from which vantage point the axes are viewed, the p_x-orbital takes no part in the bonding with this particular linear combination). The linear combination Figure 5.31 (c) matches the p_z-orbital because, like the latter, it changes sign across the xy plane, but not across the yz plane. Figure 5.31 (g) and 5.31 (e) illustrate a similar situation for the p_x- and p_y-orbitals on the carbon atom. These three are the bonding orbitals and there is an antibonding set of three corresponding to them; these are obtained by leaving the L.C.A.O. unchanged but reversing the sign of the carbon p_x-, p_y- or p_z-orbitals. When the nodes are drawn into the diagrams it is seen that molecular orbitals, Figure 5.31 (c), 5.31 (g) and 5.31 (e), have one node and are equivalent to p_x, p_y, p_z in that they are a triply degenerate set. Similarly Fig. 5.31 (d), 5.31 (f) and 5.31 (h) are a triply degenerate antibonding set.

The energy-level diagram is shown in Figure 5.32. The nodeless orbital lies lowest and the triply degenerate set, with one node, is

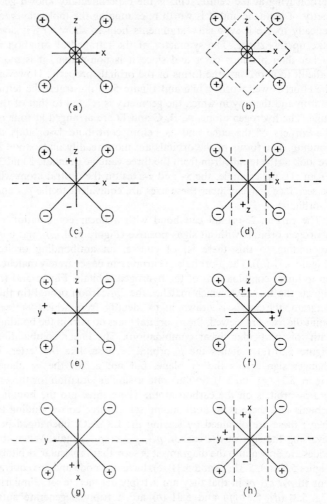

FIGURE 5.31. Orbital forms for tetrahedral methane

next in the energy scale. The antibonding orbitals turn out not to be in this simple sequence but this need not concern us now. The eight available electrons all occupy bonding orbitals and so all eight bond the molecule together, whereas it was found that in the case of planar CH_4 only six of the electrons were used in bonding. Furthermore, in the tetrahedral molecule, the nuclei and bonding electron pairs are more distant from each other and so electrostatic

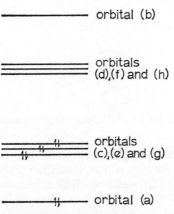

FIGURE 5.32. Correlation diagram for tetrahedral methane

repulsion would also be decreased. Taking all into account, it is not surprising that tetrahedral methane is more stable than a square planar form. This case has provided an excellent example of how in many instances it is possible to predict the geometry of a molecule directly from symmetry considerations. We would also have predicted this result from minimization of repulsive forces alone. However, it is not usually realized that the bonding forces may also be substantially altered by the symmetry of the situation.

Organometallic Compounds

As a finale to this chapter we will discuss a molecule, the existence of which, in large measure, gave impetus to the rapid development of the approach we have been exploiting. This compound is dicyclopentadienyl iron or ferrocene (Figure 5.33). For this compound, and later the isoelectronic system dibenzene chromium (Figure 5.34), no classical valency approach sufficed to explain the structure

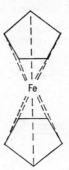

FIGURE 5.33. Structure of ferrocene

and bonding. Surprisingly, compounds similar to these had been known from as far back as 1829 when Zeise prepared a material of formula $(C_2H_4PtCl_2)_2$ which contained ethylene bonded to platinum in some way. With the advent of ferrocene and its derivatives it was not possible to shelve the bonding problems any more although it is

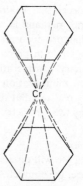

FIGURE 5.34. Structure of dibenzenechromium

not difficult to understand why chemists found it hard to comprehend the existence of such compounds. For example, Ph_2Cr contains a chromium atom which, in classical terms, is zero valent, and yet bonds with two *neutral* benzene rings! No attempt will be made to analyse in detail the molecular orbital picture of these molecules – not that this is, in principle, any more difficult than the work already accomplished, but because there are a large number of component orbitals which would obscure the essential simplicity of the problem. That the problem is similar to those previously tackled should be clear from Figures 5.33 and 5.34 because a horizontal line drawn through the central atom divides each molecule into two, just as a line through the oxygen atom divides H_2O. The *linear combination* orbitals of the rings, which are to be overlapped with those of the central atom, can be constructed from the orbital sets (themselves linear combinations) of the rings by taking 'Ring (1) orbital $\pm$ Ring (2) orbital'. (This is just like H_2 if each ring was a hydrogen atom.) Now only the π-orbitals of the rings need be considered and those of benzene have previously been studied so we could, for example, set up two linear combination orbitals of the 'two-ring' system of dibenzene chromium from the lowest π-bonding orbitals of benzene (Figure 5.16), i.e.

$$(A(1) + B(1) + C(1) + D(1) + E(1) + F(1)) \pm (A(2) + B(2) + C(2) + D(2) + E(2) + F(2))$$

$$\text{(orbital of Ring (1))} \qquad \pm \text{(orbital of Ring (2))}$$

There would, of course, be a number of others formed from the rest of the π-orbitals of benzene.

An interesting point crops up at this juncture. Is the central metal atom in such an organometallic compound just a spurious 'impurity' without which the two rings would bond together anyway? Could, e.g., there be a compound composed of two benzene rings bonded transannularly between the planes (Figures 5.35)? A bonding picture may be set up for such a 'molecule' and from it an interesting feature arises. It is found that the π-energy of the new system is exactly twice that of benzene. The explanation is that, although all twelve electrons are bonding in the 'dimer', the bonding levels are paired, one of the pair lying at a lower level than the corresponding one in benzene, whilst the other appears the same distance *higher* in the energy scale (Figure 5.36). Thus no overall stabilization can result from the association of the two rings. It seems, therefore, that the presence of the central metal atom is imperative for formation of a stable complex. The situation is rather similar to that we found when examining the saturation phenomenon.

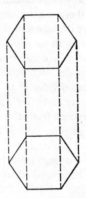

FIGURE 5.35. Benzene 'dimer'

The final organometallic compounds are described by combining the 'Ring 1 $\pm$ Ring 2' linear combinations with orbitals of the same assigned symmetry on the central atom. In transition elements the latter will be s, p and d type.

Using the techniques we have outlined in this chapter, one can qualitatively examine many actual or model molecular situations, and in this way learn a great deal about the structure of the system – indeed this is frequently definitive.

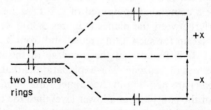

FIGURE 5.36. Complementary levels of the benzene 'dimer'

PROBLEMS FOR CHAPTER 5

1. Why does the NO_2 group exert mainly an inductive and not a mesomeric effect when substituted into a benzene ring? (Hint: study the energy-level diagram and consider the a_2 and b_2 orbitals.)

2. The allyl radical $[CH_2{=}CH{-}CH_2]^{\cdot}$ is angular and possesses an odd π electron. Construct the π-orbital forms and energy-level diagram for the system and its associated cation and anion.

   What do you notice about the energies of electron jumps in these systems? (Hint: refer to NO_2.)

3. Set up an energy-level diagram for NeH_2.

4. Construct the π-energy-level pattern of trimethylene methane, i.e.

   (Hint: refer to BH_3 and NO_2.)

5. What would be the orbital pattern of H_4 if (a) one hydrogen were surrounded by three equivalent ones, (b) each H atom was situated at the corner of a regular tetrahedron?

6. Show that the linear combinations,

$$(p_A - p_B) \pm (p_C + p_D)$$

   are not correct wave functions for the π-orbital system of butadiene.

CHAPTER 6

Summing-up

In this final chapter the ideas of Chapter 1 and the final picture of chemical bonding will be compared. It is necessary therefore first to recapitulate briefly on Chapter 1, and then to sum up Chapters 2, 3, 4 and 5.

We started with the idea that an integral number, the valency, might be assigned to an atom and, by virtue of this, the atom could form a number of bonds to other atoms or groups. Two main kinds of bonding were next recognized; these were called ionic and covalent bonding. The first stemmed from simple mutual electrostatic attraction of charged units, whilst in binary compounds the second was considered to result from the pairing of electrons, one from each atom. Rigid application of these ideas led to a number of problems associated with the combining power of atoms, particularly in inorganic compounds. What we have tried to do in this book is firstly to point out the difficulties inherent in the dogma that valency is a property of the atom and can be completely defined, and secondly to develop a more general approach to the problem of molecular bonding.

The theory of molecular bonding was developed *via* orbital theory. Firstly the behaviour of electrons in atoms was described in terms of atomic orbitals, emphasis being placed on the substitution of electron density or probability for the fixed Bohr atomic orbits. Simple mathematical properties of atomic orbitals were next studied, their energy interrelationships, the significance of their nodes, and the interpretation of the constituent angular and radial and the derived radial distribution functions. The orbitals were then filled up from the lowest in energy and hence the electronic structures of the lighter elements in the Periodic Table were rationalized.

The interrelationship of ionic and covalent bonds was readily derived by analysing the attractive and repulsive forces inherent in molecules. Next two ways in which bonding can be discussed, the valence bond and the molecular orbital theories, were considered and it was stressed that the theories only describe molecules. The valence-bond theory is a theoretical expression of the electron-pair bond ideas of Kossel, Lewis and Langmuir and demands that orbitals taking part in simple covalent bonding must contain unpaired electrons. When this condition does not initially obtain, formal promotion of electrons to empty orbitals and hybridization are invoked to yield a description of the system.

Mutual overlap of atomic orbitals on different centres was found to be an important factor in determining the occurrence of chemical bonding – the way in which two orbitals can overlap is governed by their relative orientation and is frequently zero. A study of orbital overlap led to the concept of two different kinds of bond, σ and π, and these were illustrated by reference to unsaturated organic hydrocarbons. For some compounds the contribution of resonance hybrids to the overall picture explained their physical properties. The latter addend represented a considerable advance on the simple electron-pair idea because it explained both odd-electron systems like nitric oxide and the observed equivalence of all the bonds in $SO_4^=$ and benzene. In the latter application it led to the idea of an 'average' bond between the atoms. Extension of the valence-bond treatment to heteropolar systems showed that they may be described by orbital overlap in a similar way as was employed for homopolar systems. This led to the important conclusion that there is essentially no difference in the valence-bond approach to the formulation of an ionic or a covalent bond. The form of the overall wave function for the system reflects the disparity in the electron distribution between the two atoms, and so all transitional stages between 'pure' covalent and 'pure' ionic bonds are adequately incorporated under one unified treatment.

The second approach to chemical bonding, molecular orbital theory, started from different principles, i.e. unpaired electrons in overlapping orbitals were not required in order to bring about bonding. This particular facet of the theory has its strength in that it can deal equally well with systems amenable to the electron-pair approach and with those which are out of line with this requirement. The basic tenet of m.o. theory is that the symmetry of the system under discussion is used in order to set up the orbitals for the molecule. Their energy sequence is then obtained. Having done this, the available electrons are fed into the orbital levels so generating the bonding pattern for the molecule.

Since there is no initial assumption of unpaired electrons on the constituent atoms, a rather broader view of valency emerges than from the valence-bond approach. One need not *a priori* assume that a given atomic pattern must necessarily be unstable simple because there are insufficient electrons to assign to each pair of atoms or because the ground state of one atom has closed electron shells. On the other hand, addition of further electrons to a system to complete pairs does not necessarily strengthen the bonding, e.g. the bonding in the nitric oxide molecule is weakened by adding another electron.

Each postulated model is investigated on its merits and the disposition of electrons throughout the system discovered. The electron occupancy of individual bonds results from contributions from the whole set of filled orbitals. No trouble arises with unpaired electrons as these are fed into the level scheme just as are paired ones. The concept of 'valency' as meaning the number of bonds that an element can form is very often completely lost because the number varies from compound to compound. The molecular orbitals encompass the whole molecule and each atom is, to some extent, bonded to all the others. In these cases it becomes quite irrelevant to try to assign a valency number to the atom. This is particularly true of atoms in inorganic compounds. This was the tentative conclusion to which we were propelled in Chapter 1 – the number of anomalies made it pointless to attribute a valency number to many atoms. The answer to the questions of valency raised in that chapter with respect to NaCl, H_2O and NH_4^+ is that we do not try to answer them, because no 'answer' would really get us any further forward.

We can, of course, continue to take the best from the older approach. Indeed the application of symmetry arguments will frequently lead to the same overall answers and the generality introduced by the latter will not be required. It is imperative that, at all stages, the wood is not obscured by the trees. The present argument is that we will be better prepared to understand the whole of chemistry if we are thoroughly conversant with the more general principles.

Oxidation Numbers

It has been shown that to assign an integral number to the valency of an atom is fraught with difficulty. It is, however, quite useful to define the *oxidation state* of an element in a compound and this is an integral quantity. The method of arriving at the *oxidation number* is usually unambiguous. A simple way of determining its value is to calculate the charge remaining on the ion when the groups attached to it are completely removed together with their pairs of electrons. Thus in an isolated sodium chloride molecule the oxidation number of both sodium and chlorine is unity under this definition because removal of Na^- or Cl^- leaves Cl^+ or Na^+.

In methane, CH_4, the oxidation number of carbon is four and in borine, BH_3, that of boron is three. In the nitrate anion, NO_3^-, removal of the three oxygen atoms with their three pairs of electrons leave a charge of $(6^+)^-$ on nitrogen, i.e. 5^+, which is the oxidation number of nitrogen in this ion. Notice here that the oxidation number is a well-defined quantity whereas the number of bonds formed by nitrogen in NO_3^- would be rather more difficult to define.

What Next?

In this book we have attempted to provide simple reasons why chemical bonds are formed and to formulate ways of describing them. Both these ends can be achieved in an elementary way. The question now arises: What is the next stage in the study? In *qualitative* terms there is not much more that need be done, the rules outlined here will suffice for most applications. The symmetry arguments can, of course, be refined technically for cases of high symmetry.

However, *quantitatively*, there is much further to go. Apart from reference to the nodes in wave functions serving as a guide to the energy sequence, there was no real attempt at any stage in the present book to put any of the subject matter on a quantitative basis. What we should now do is to redevelop the quantitative theory of molecular energy levels and to show how the theory can be implemented with respect to any system. We would then be able to apply the results to a wide area of chemistry in cases where qualitative argument based on symmetry was insufficiently definitive.

It is not difficult to proceed further and apply symmetry concepts to the determination of the forms of various molecular spectra. This is of immense practical importance in spectroscopy but is outside the scope of the present volume.

Suggestions for Further Reading

(1) C. A. Coulson, *Valence*, Oxford University Press, London.
(2) J. N. Murrell, S. F. A. Kettle and J. M. Tedder, *Valence Theory*, Wiley, New York.
(3) F. A. Cotton, *Chemical Applications of Group Theory*, Wiley, New York.

Index